Horticulture: Principles and Practices

Horticulture:
Principles and Practices

Edited by **Thelma Bosso**

New York

Published by Callisto Reference,
106 Park Avenue, Suite 200,
New York, NY 10016, USA
www.callistoreference.com

Horticulture: Principles and Practices
Edited by Thelma Bosso

International Standard Book Number: 978-1-63239-420-0 (Hardback)

Printed in the United States of America.

Contents

Preface   VII

Part 1   Soil Management   1

Chapter 1   **Influence of Fertilizers with Prolongation Effect on Productivity of Root-Crop Vegetables and Biochemical Composition Before and After Storage**   3
Ona Bundinienė, Danguolė Kavaliauskaitė, Roma Starkutė, Julė Jankauskienė, Vytautas Zalatorius and Česlovas Bobinas

Chapter 2   **Effect of the Climate and Soil Characteristics on the Nitrogen Balance in the North of Algeria**   25
N. Bettahar

Part 2   Plant Breeding   41

Chapter 3   **Assessment of Diversity in Grapevine Gene Pools from Romania and Republic of Moldova, Based on SSR Markers Analysis**   43
Ligia Gabriela Gheţea, Rozalia Magda Motoc, Carmen Florentina Popescu, Nicolae Barbacar, Ligia Elena Bărbării, Carmen Monica Constantinescu, Daniela Iancu, Tatiana Bătrînu, Ina Bivol, Ioan Baca and Gheorghe Savin

Part 3   Protected Horticulture   61

Chapter 4   **Total Growth of Tomato Hybrids Under Greenhouse Conditions**   63
Humberto Rodriguez-Fuentes, Juan Antonio Vidales-Contreras, Alejandro Isabel Luna-Maldonado and Juan Carlos Rodriguez-Orti

Part 4   Postharvest Physiology   73

Chapter 5   **Chemical Composition and Antioxidant Activity of Small Fruits**   75
Pranas Viskelis, Ramune Bobinaite, Marina Rubinskiene, Audrius Sasnauskas and Juozas Lanauskas

Part 5   **Urban Horticulture**   103

Chapter 6   **Urban Horticulture and Community**
            Economic Development of Lagging Regions   105
            Albert Ayorinde Abegunde

Chapter 7   **Local Botanical Knowledge and Agrobiodiversity:**
            Homegardens at Rural and Periurban
            Contexts in Argentina   117
            María Lelia Pochettino, Julio A. Hurrell and Verónica S. Lema

Part 6   **New Technologies**   145

Chapter 8   **Postharvest Technologies**
            of Fresh Horticulture Produce   147
            Alejandro Isabel Luna-Maldonado,
            Clement Vigneault and Kei Nakaji

Chapter 9   **Sustentable Use of the Wetting Agents**
            in Protected Horticulture   159
            Carlos Guillén and Miguel Urrestarazu

            Permissions

            List of Contributors

Preface

This book aims to highlight the current researches and provides a platform to further the scope of innovations in this area. This book is a product of the combined efforts of many researchers and scientists, after going through thorough studies and analysis from different parts of the world. The objective of this book is to provide the readers with the latest information of the field.

This book illustrates the principles and practices of horticulture with the help of up-to-date information. It discusses modern features and future trends in this field. The various topics presented in this book include the effect of climate and soil properties on nitrogen balance, impact of fertilizers with prolongation effect, variedness in grapevine gene pools, growth and nutrient intake of tomato plants, post-harvest quality, chemical composition and antioxidant activity, local botanical information and agrobiodiversity, urban horticulture, utilization of humectant agents in protected horticulture and post-harvest techniques of fresh horticulture produce. This book will cater as a valuable source of reference for students, professional horticulturalists and readers interested in acquiring knowledge about this field.

I would like to express my sincere thanks to the authors for their dedicated efforts in the completion of this book. I acknowledge the efforts of the publisher for providing constant support. Lastly, I would like to thank my family for their support in all academic endeavors.

Editor

Part 1

Soil Management

Influence of Fertilizers with Prolongation Effect on Productivity of Root-Crop Vegetables and Biochemical Composition Before and After Storage

Ona Bundinienė, Danguolė Kavaliauskaitė, Roma Starkutė,
Julė Jankauskienė, Vytautas Zalatorius and Česlovas Bobinas
Institute of Horticulture, Lithuanian Research Centre for Agriculture and Forestry,
Lithuania

1. Introduction

Because of the increasing requirements to environment protection and safe food it is necessary to increase the use of non-toxic, environment-friendly substances (Malekian et al., 2011). Vegetable fertilization must correspond to the requirements of ecology, i. e., it is necessary to fertilize environment less and to increase plant productivity and protection. We must look for the suitable and optimal fertilization methods and fertilizer types. Growing foliage needs a lot of nitrogen and root-crops – potassium and phosphorus. Calcium participates in plant growing, nutrient metabolism and many biochemical and physiological processes (White et al., 2003; Saure, 2005). Its deficit in reproductive tissues worsens the quality (Grattan & Grieve, 1998; Porro et al., 2002). The lack of magnesium most often is observed in acid, light soils. Vegetable production storability is better when it is fertilized with potassium and calcium at the end of vegetation. Moreover, calcium stops soil acidification processes (Liet. dirvožemių ..., 1998). It is possible to satisfy the necessity of calcium and to prevent soil acidification by fertilizing with physiologically nonacid fertilizers (Аутко, 2004). This may be ammonium nitre, in the composition of which there are 27 % N (by 13,5 % ammonium – $N-NH_4^+$ and nitric – $N-NO_3^-$), 6 % calcium (CaO) and 4 % magnesium (MgO) or nitrogen fertilizer with zeolite, in the composition of which there are 26 % N and a part of dolomite (up to 6 %) is substituted with zeolite. Besides, there are 4.5–5.4 % of calcium (CaO) and 3.1–3.5 % of magnesium (MgO). Calcium and magnesium are extracted from natural dolomite. They decrease ammonium nitre physiological acidity and thus balance soil acidity and improve its biological activity. Insertion of zeolite into calcium ammonium nitre granule improves fertilizer's physical properties, its friability, also decreases the wash out of nutrients, especially ammonium nitrogen, potassium and calcium and improves solubility of non-soluble combinations (Butorac, 2002; Ambruster, 2001; Ramesh et al., 2011; Uher, 2004; Yolcu et al., 2011). According to the data of many various counties (Li et al., 2002; Polat et al., 2004; Gül et al., 2005), sorption properties of zeolite guarantee 15–30 % more economical use of nitrogen ant increase yield, prolong the duration of nutrient effect and decrease the necessity of often fertilization, especially in the soils, which lack nutrients (Lobova, 2000; Beqiraj (Goga) et al., 2008). Moreover, fertilizing with granular fertilizers most nutrient elements remain in 5-10 cm soil layer. Executing EU Council directive 91/676/EEB about water protection from the pollution by nitrates used in agriculture (Nitrates Directive), there are looked for the possibilities to improve nitrogen fertilizers, improving the conditions of plant nutrition. Rines et al. (2006; Rehakova et al., 2004) states that zeolite is the suitable means for creating good conditions for plant growing, being satisfied with less amount of nutrients and water. In 2003 there was started to produce fertilizer with zeolite, which the main compound is klinoptilolite (Ancuta et all., 2011; Mažeika et al., 2008).

The possibilities of different nitrogen fertilizers ((calcium ammonium nitre (CAN 27) N_{90+30}), nitrogen fertilizer with zeolite (N 26 + 6 % zeolite) and zeolite (commercial sign ZeoVit EcoAgro) 2,5 kg m^{-1}) and ammonium nitre (AN N_{90+30})) have been investigated.

2. Materials and methods

2.1 Agrochemical properties of soil

Experiments were carried out in calcaric epihypogleyic luvisol of sandy loam on light loam soil *Calcari – Epihypogleyc Luvisols (LVg-p-w-cc)*. Soil, in which red beet and carrot were grown, was little nitric (in lay of 0–60 cm 45,0 –48,8 kg ha^{-1}), rich in agile phosphorus (364,4–411,8 mg kg^{-1}), calcium (7941–8638 mg kg^{-1}) and magnesium (1952–2357 mg kg^{-1}), averagely rich in agile potassium (154,1 – 165,1 mg kg^{-1}), and there was little amount of humus in it (1,40–1,54 %). Soil pH 7,2 – 7,6.

2.2 Taking and storage of samples

Hybrid red beet 'Pablo' F_1 (500 thousand unt. ha^{-1} of germinable seeds) was grown on flat surface, carrot cultivar 'Samson' (800 thousand unt. ha^{-1} of germinable seeds) – on furrowed surface. Sowing scheme – 62 + 8. Work of plant supervision was carried out according to vegetable growing technologies accepted in LIH. Red beet and edible carrot yield was gathered, when vegetables reached technical maturity. When harvesting according to the variants, with three replications, there were taken samples for biochemical investigations and 12–15 kg samples for root-crop storability investigations. Root-crops were stored in freezer, under stabile temperature (-1- +2 °C) and relational humidity (85–90 %), in the storage houses of Institute of Horticulture, LRCAF, Biochemistry and Technology laboratories. Root-crop storability was inspected after 3 and 7 months, classifying well-preserved and diseased (rotten, partly rotten and wilted, i. e. not suitable for usage) red beet root-crops and establishing the natural loss, i. e., drying.

Soil samples for the investigations of agrochemical properties were taken in autumn, after yield gathering, when the jointed sample according to the variant was created.

2.3 Methods of analysis

Investigations of red beet biochemical composition were carried out at the Laboratory of Biochemistry and Technology, Institute of Horticulture, LRCAF. There was established: dry matter – gravimetrically, after drying out at the temperature of 105 ± 2 °C up to the unchangeable mass (Food analysis, 1986), dry soluble solids – with refractometer (digital

Influence of Fertilizers with Prolongation Effect on Productivity of Root-Crop Vegetables
and Biochemical Composition Before and After Storage

5

refractometer ATAGO) (AOAC, 1990a), sugars – by AOAC method (AOAC, 1990b), nitrates – by potentiometrical method with ion selective electrode (Metod. nurod., ...1990), carotenoids – spectrophotometrically (Davies, 1976).

Investigations of soil agrochemical composition were carried out in the center of Agrochemical investigations LAI (now Center of Agrochemical Investigations of LRCAF).

There were established soil agrochemical indices: pH_{KCl} – ISO 10390:2005 (potenciometrical); agile P_2O_5 and K_2O – GOST 26208-84 (Egner-Riehm-Domingo – A-L), humus – ISO 10694: 1995 (dry burning), calcium and magnesium – SVP D-06 (atomic absorption spectrometrical), mineral nitrogen – 1M KCl extraction, spectrophotometrical MN-1984.

2.4 Growing cultivars, scheme of experiments, applied fertilizers

Hybrid red beet 'Pablo' F_1 (500 thousand unt. ha^{-1} of germinable seeds) was grown on flat surface, carrot cultivar 'Samson' (800 thousand unt. ha^{-1} of germinable seeds) – on furrowed surface. Work of plant supervision was carried out according to vegetable growing technologies accepted in LIH.

Scheme of the experiment:

1. Without N, Ca, Mg, PK before sowing – background (B – $P_{60}K_{120}$)
2. B + calcium ammonium nitre N_{90} before sowing + N_{30} at the stage of 4 – 6 leaves (B + CAN 27 $N_{90 + 30}$)
3. B + calcium ammonium nitre with zeolite N_{90} before sowing + N_{30} at the stage of 4 – 6 leaves (B + N 26 + 6z $N_{90 + 30}$)
4. B + ammonium nitre N_{90} before sowing + N_{30} at the stage of 4 – 6 leaves + zeolite (B + AN $N_{90 + 30}$ + zeolite).
5. B + dusting with zeolite* (i. e., root-crops were sprinkled and mixed with zeolite). Root-crop samples (15 kg each) were taken from the first variant and sprinkled with zeolite 30 kg t^{-1}.

For the background fertilization there was used granular superphosphate and potassium sulphate. Nitrogen fertilizers were applied in the rates and forms indicated in the scheme. In the last variant before vegetable sowing there was inserted 2,5 kg m^{-1} of zeolite. Experiments were carried out every year in 4 replications in randomized fields, storage investigations – in 3 replications (10–12 kg of vegetables per each) storing them in polypropylene bags. Storing vegetables, variant with zeolite was added, vegetable samples were taken from background fertilization (the first variant) and sprinkled (felted up) with zeolite 30 kg t^{-1}. Area of record plot – 6,2 m^2. the analyses of vegetable biochemical composition and soil agrochemical indices were carried out in 3 replications.

2.5 Mathematical procession

Data significance was evaluated by one-factorial dispersion analysis, using program ANOVA; the relation among different indices – by correlation-regression analysis; using program STAT_ENG. Using indicises: r – coreliation coefficient, ** – level of probability 01, * – level of probability 05.

2.6 Meteorological conditions

Meteorological conditions in different years of investigation were different: vegetation period in 2004 was cooler than multiannual and the amount of precipitation bigger than multiannual average, 2005 was warm and dry, 2006 – hot and humid, 2007 – cool and humid (table 1). Especially many precipitations fell in August of 2004, 2005 and 2006. Air temperature was very different. In 2004 and 2006 it was much higher than the multiannual average, but in 2006 it was even in 2 °C cooler. July of 2005 was especially dry and hot. In July 2006 precipitations comprised only 40 % of multiannual rate ant the temperature was in 2 °C higher than the multiannual average.

Month	2004	2005	2006	2007	Multi annual	2004	2005	2006	2007	Multi annual
May	11,1	11,3	12,6	11,2	12,3	46,2	65,4	74,0	104,4	50,7
June	13,7	14,9	16,3	15,1	15,9	77,4	66,6	13,8	72,2	71,2
Jule	16,4	19,1	19,3	15,2	17,3	50,4	3,8	30,2	173,6	75,3
August	17,3	14,7	17,5	16,6	16,7	118,2	109,4	173,4	42,8	78,4
September	10,1	14,2	14,5	10,6	12,1	36,2	46,5	83,0	57,8	58,7
Average	13,7	14,8	16,0	13,7	14,9	65,7	58,3	74,9	90,2	66,9

Table 1. Meteorological conditions during vegetation period. Data of Babtai agrometeorological station, iMETOS prognostication system

3. Results and discussion

3.1 Yield of red beet and carrot

Data of investigations carried out in various countries with various plats show that nitrogen is the factor, which determines the growth and productivity of plant most of all (Scholberg et al., 2000; Babik & Elkner, 2002; Tei et al., 2002; 2001; Rubatzky et al., 1999; Malnou et al., 2008). Nevertheless, nitrogen induced environmental damage such as nitrate pollution and wasting fossil fuel (Fustec et al., 2009). Red-beet productivity, independently of their cultivars and types, also increases applying nitrogen fertilizers (Ugrinovic, 1999; Staugaitis & Tarvydienė, 2004). According to our data of investigations carried out at the Institute of Horticulture, LRCAF, in 2004–2007, red beet yield increased averagely 2,1 times (Fig. 1), using nitrogen fertilizers (N_{90+30}), independently from fertilizer form, in comparison with the yield obtained when red-beet were fertilized only with phosphorus and potassium fertilizers ($P_{90}K_{120}$). The yield of edible carrot increased 14 % (Fig. 2). The output of marketable red-beet yield increased averagely 19,0 %, carrot – 3,5 %. The output of marketable yield is important parameter, which determined fertilizer suitability and corresponds to one of the main requirements of the optimal yield – high output of marketable yield (Suojala, 2000; Zalatorius & Viškelis, 2005). Marketable red-beet yield after fertilizing with ammonium nitre increased by 22,6 t ha[-1] or in 93,4 % in comparison with this of red-beet grown without nitrogen fertilizer; after fertilizing with ammonium nitre plus zeolite – 28,7 t ha[-1] or 2,1 times. There was obtained the bigger marketable yield after fertilizing with ammonium nitre and inserting zeolite, because the output of marketable yield was better in 3,3 %. Fertilizing carrots both with ammonium nitre and ammonium nitre with zeolite there was obtained equal yields. Using nitrogen in both variants, marketable carrot yield increased 6,0 t ha[-1] or 12,3 %; fertilizing with ammonium nitre and zeolite the output of marketable yield increased 2,0 %. The data of foreign scientists show that fertilizing with calcium ammonium nitre the yield increases. When malty barley was fertilized with calcium ammonium nitre, dependently on soil granulometrical composition, the biggest additional yield was obtained after scattering 112.5–137,5 kg ha[-1] N (Conry, 1997). According to the data of Zdravkovic et al. (1997), carrot fertilized with manure produced 48,4 t ha[-1], with calcium ammonium nitre – 41,5 t ha[-1] and mixture – 41,5 t ha[-1] of yield. Sorbtional properties of zeolite, according to the data of many investigators (Challinor et al. 1995; Ilsildar, 1999; Li et al., 2002; Polat et al., 2004), guarantee 15–30 % more economical nitrogen usage, prolong the duration of nutrients action and decreases the necessity of often fertilization. Natural zeolites are nature's own slow release fertilizers (Li, 2003; Середина, 2003).

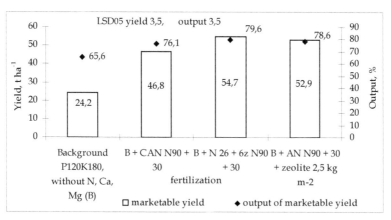

Fig. 1. Influence of fertilizers with prolongation effect on productivity of red beet. Babtai, 2004–2007

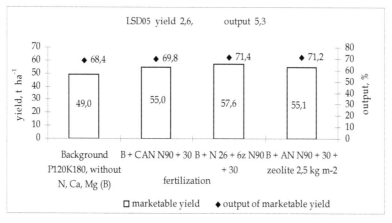

Fig. 2. Influence of fertilizers with prolongation effect on productivity of edible carrot. Babtai, 2004–2007

Grecian scientists (Samartzidis et al., 2005) state that zeolite doesn't have positive influence on rose productivity. Zeolite efficiency is influenced by the size of granules. 3–5 mm zeolite granules are of the biggest sorption susceptibility. According to Russian investigators (Ryakhovskaya & Gainatulina, 2009), the use of zeolites allows reducing the basic fertilizer dose by 25 % and growing annual grasses without fertilizers in the second and thirds years after it application. Introduction of zeolite in our investigations positively influenced the increase of marketable yield output. The data of Siberian peat investigations show that the presence of zeolite in granule, when fertilizing in all directions, didn't produced additional yield, but fertilizing locally increased yield 9–13 %, in comparison with yield, obtained fertilizing only with the granular mixture of peat-mineral fertilizers (Алексеева et al., 1999). The data of Polish investigators showed that nitrogen fertilizer with DMPP nitrification inhibitor is suitable as nitrogen fertilizer, equal to ammonium and calcium nitrates and more effective than ammonium sulphate (Kolota et. al., 2007). The data of the carried out experiments showed that the biggest red beet and carrot yields and the bigger output of the marketable yield, than this obtained fertilizing with other nitrogen fertilizers, was obtained applying nitrogen fertilizer with zeolite (N 26 + 6c N_{90+30}).

3.2 Storage of red beet and carrot crop-root

Vegetable storability is influenced by climatic conditions, soil, cultivars, fertilization, forms of fertilizers and the time of harvesting (Suojala, 2000; Sakalauskas et al, 2004; Rożek et al., 2000). In order to preserve vegetables it is important to keep the suitable temperature, humidity and to create suitable conditions for breathing (Raju et al., 2010; Workneh et al., 2011; Badelek et al., 2002; Kołota et al., 2007). One of the most valuable red beet farm properties is uncomplicated their growing, good biochemical composition and good storability (Petronienė & Viškelis, 2004). Important property of red beet, which improves their storability, is their thicker skin and root-crop ability to pass to the state of tranquility (Аутко, 2004). Round root-crops of red beet is stored better than cylindrical root-crops. The output of marketable production, when growing red beet without nitrogen fertilizers after short-time storing (3 months) comprised 82,6 %, after long-time storing (7 months) – 59,5 %, and losses correspondingly – 17,4 % and 40,5 % (Fig. 3). Additional fertilizing with nitrogen fertilizers the output of marketable red beet production after short-time storing increased averagely 6,3 %, after long-time storing – 21,7 %. The amount of marketable production, suitable for realization increased correspondingly 19,1 t ha-1 or 94,6 % and 17,3 t ha-1 or 2,2 times. Red beets, fertilized with nitrogen fertilizer with zeolite, were preserved best of all both after 3 and 7 months. The amount of marketable production after 3 months storage, in comparison with this one of red beet fertilized with calcium ammonium nitre, increased 7,3 t ha-1 or 17,7 %; in comparison with this one of red beet fertilized with ammonium nitre with zeolite – 2,3 t ha-1 or 5,1 %. After long-time storage the amount of marketable production increased correspondingly 8,4 t ha-1 or 26,5 % and 3,3 t ha-1 or 9,1 %. Red beet root-crops sprinkled with zeolite were stored very well. After short-time storage the output of marketable production, in comparison with this one of red beet grown without nitrogen fertilizers, increased 8,6 %, i. e., storing the same amount of marketable yield as in the background variant there was obtained 1,5 t ha-1 more marketable production. After long-time storage the output of marketable production increased 35,5 %, i. e., it was obtained 4,7 t ha-1 more of marketable production than storing red beet root-crops fertilized only with phosphorus and potassium fertilizers. It is thought that red beet root-crop storability was improved by zeolite ability to hold up humidity.

To store carrot is more difficult than other root-crop vegetables. It is very important for them temperature and humidity during storage (Suojala, 2000; Fikseliova et al., 2010). Carrot root-crop have thin cover tissue (4–8 layers of periderma, when this one of potatoes – 9–11 layers), which during yield gathering with mechanical means very often is injured. That is why water is evaporated more intensively (Аутко, 2004). Carrots quickly wilt, and wilted are less resistant to diseases, but small mechanical injures root-crop is able to "heal up". Meteorological conditions during the last two weeks before gathering have big influence on carrot storability (Fritz & Weichmann, 1979). Our data showed, that storing carrot for short time (up till New Year) marketable production comprised averagely 46,6 t ha⁻¹, and after long-time storage (up till May) – 38,5 t ha⁻¹ (Fig. 3). Storage losses correspondingly were 12,3 % and 27,4 %, i. e., carrots, suitable for realization comprised averagely 87,7 % and 72,6 %.

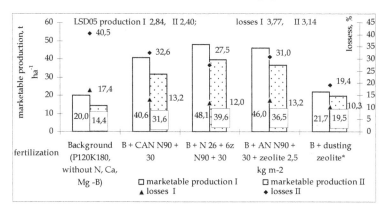

Fig. 3. Influence of fertilizers with prolongation effect on amount marketable production and persistence of red beet crop-root.

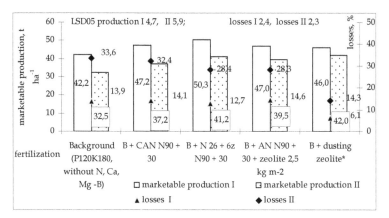

Fig. 4. Influence of fertilizers with prolongation effect on amount marketable production and persistence of carrot crop-root. Babtai, 2004–2008.

The amount of carrot, fertilized with nitrogen fertilizers, marketable production after short-time storage increased on the average 5,4 t ha⁻¹ or 12,9 %, after long-time storage– 7,5 t ha⁻¹ or

23,0 %. The results of storage investigations showed that carrots, fertilized with nitrogen fertilizer with zeolite, were preserved best of all. The amount of marketable production after short-time storage, in comparison with this one of carrot grown without nitrogen fertilizers, increased 8,1 t ha-1 or 19,2 %. Storing carrot till spring (7 months), storage losses decreased from 33,6 % up to 28,4 % and it was obtained additionally 8,7 t ha-1 carrot marketable production. Carrot root-crop sprinkling with zeolite very improved their storability (from 86,1 % to 93,9 %, and after long-time storage from 66,4 % to 85,7 %). The amount of marketable production, in comparison with this one of carrot grown without nitrogen fertilizers, increased correspondingly 3,8 t ha-1 or 9,0 % and 9,5 t ha-1 or 29,2 %.

3.3 Quality of red beet and carrot crop-root

Mineral fertilization is one of the most important and effective factors influencing metabolism and at the same time yield quality. Plants assimilate nutrients, which they receive with mineral and organic fertilizers (Lairon, 2010). Their concentration in vegetables changes dependently on many factors: vegetable type and cultivar (Montemurro et al., 2007; Rożek et al., 2000), soil, meteorological conditions (Rubatzky et al., 1999; Suojala, 2000; Jalali, 2008). Too intensive fertilization, especially with nitrogen, can cause unsuitable increases in some plants of nitrates, sugars and decreases of dry soluble solids, ascorbic acid (vitamin C), calcium and magnesium (Wang et al., 2008; Sorensen, 1999), therefore, it is very important not to delay fertilization and to use suitable fertilizers (Petronienė & Viškelis, 2004). Fresh red beet root-crops accumulate 16–22 % of dry matter, 10–16 % of sugars, 9–32 mg% of vitamin C, small amounts of other vitamins (B_1, B_2, PP), and the color of root-crop depends on the amount of betain (Аутко, 2004). According to the data of investigations carried out in Lithuanian, meteorological conditions, cultivar and soil influence red beet root-crop biochemical composition more than fertilizers (Staugaitis & Dalangauskienė, 2005; Tarvydienė & Petronienė, 2003; Butkuvienė et al., 2006). Lithuanian investigators (Petronienė & Viškelis, 2004a) indicate that the amounts of dry soluble solids in red beet root-crops can be 8,3–16,2 %, these of sugars – 4,98–12,6 %, ascorbic acid – 9,0–31,2 %, betanins – 39,9–96,7 mg 100 g-1, nitrates – 272–2322 mg kg-1. Our investigations showed that after yield gathering the amounts of dry matter and dry soluble solids in red beet root-crop growing them without nitrogen fertilizers were correspondingly 14,4 % and 12,3 % (Fig. 5).

The least amounts of dry matter and correspondingly dry soluble solids were found in root-crops, fertilized with nitrogen fertilizer with zeolite (11,9 and 11,7 %). Precipitation positively influenced the amounts of dry matter ($r = 0,73^{**}$), dry soluble solids ($r = 0,69^{**}$) and sugars ($r = 0,83^{**}$) and decreased the amount of nitrates ($r = -0,59^*$). Temperature effect was opposite (correspondingly $r_{dm} = -0,73^{**}$, $r_{dss} = -0,64^{**}$, $r_{sugars} = -0,75^{**}$, $r_{nitrates} = 0,60^*$). In the stored vegetable, similarly as during the growth, constant metabolism takes place. Vegetable storage and nutrient losses depend on its intensity. When vegetables are stored in low temperature and suitable humidity, breathing and all the biochemical processes slow down. The most suitable temperature for the storage of root-crop vegetables is from -1 to +2 °C, relational humidity – 85–95 % (Аутко, 2004; Petronienė & Viškelis, 2004). According to the data of Polish investigators (Badelek et al., 2002), the least amount of non marketable red beet root-crop and the best their quality are when in storing houses +2 °C temperature is kept and the size of red beet doesn't have influence on the storage. According to the data of German investigators (Henze & Bauman, 1979), humidity during storage has bigger influence on red beet root-crop preservation than temperature. Red beets are stored better when relational humidity is more than 95 %. Data of our investigations show that during red beet storing the amounts of dry matter in root-crops, in comparison with amounts during yield gathering, decreased fertilizing with nitrogen fertilizer with zeolite and storing red beet, grown without nitrogen fertilizers, sprinkled with zeolite; it increased fertilizing with calcium ammonium nitre and ammonium nitre with zeolite. The amounts of dry soluble solids in all the fertilization variants increased. According to Lithuanian investigators (Karklelienė et al., 2007), the amounts of dry soluble solids after storage depend on genetype. Red beet cultivar 'Kamuoliai 2' was distinguished for the bigger amount of dry soluble solids. Genetype also influence the amount of sugars in red beet root-crops (Karklelienė et al., 2009). The amount of sugars, according to Petronienė & Viškelis (2004a), influences root-crop nutritional properties and procession. Red beets, which have more sugars, are distinguished for better taste properties. Growth conditions influence the amount of sugars too. In the investigations the amounts of sugars didn't fluctuate in wide limits (Fig. 6).

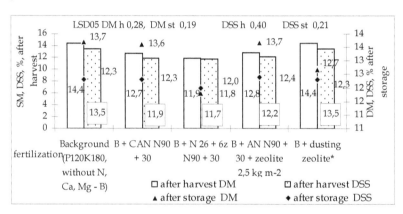

Fig. 5. Influence of fertilizers with prolongation effect on content of dry matter and dry soluble solids in red beet crop-root. Babtai, 2004–2008.

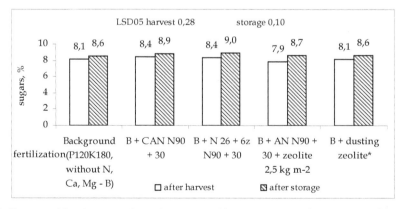

Fig. 6. Influence of fertilizers with prolongation effect on content of sugars in red beet crop-root. Babtai, 2004–2008.

Fertilizing with nitrogen fertilizers the amounts of sugars decreased insignificantly (on the average 0,3 % unt.). They were equal both fertilizing with calcium ammonium nitre and nitrogen fertilizer with zeolite. The amount of sugars depended on the amounts of dry matter (r = 0,50*) and dry soluble solids (r = 0,57**). After storage the amounts of sugars in red beet root-crops slightly increased (from averagely 8,1 % to 8,7 %). The biggest amount of them was in red beet root-crops fertilized with nitrogen fertilizer with zeolite. Polish investigators (Rożek et al., 2000) indicate that during root-crop storage the amounts of soluble sugars and nitrated decrease, but the amounts of phenols increase. Our data showed that sugar amount negatively correlates with nitrate amounts (r = -0,56*) (Fig. 7). Klotz et al. (2004) indicated that sugar beet sucrolytic activities change little during storage, regardless of storage temperature, length of storage. Polish investigators (Szura et al., 2008) indicate that the type of nitrogen fertilizer doesn't influence the amounts of dry matter, sugar, phenols and ammoniac nitrogen (NH_4) and proteins, phosphorus, potassium and magnesium, and that fertilizes with nitrification inhibitor (Entec 26) decreases nitrate amounts. Nitrates, as Lairon (2010) states, are absorbed through roots and naturally accumulate in plants, where later on are used for amino acid synthesis. Even fertilizers rich in nitrogen, especially of organic origin, when there are high soil mineral level, do not accumulate much nitrates and their accumulation also depend on meteorological conditions, plant cultivars and yield gathering time. Some scientist affirm that nitrate amount increases when nitrogen amount increases in the soil, plants suffer stress (shade, drought, etc.) and additionally fertilizing with leaf fertilizer or applying combined fertilization on soil surface and through leaves (Alexandrescu et al., 2000). The data of statistic analysis of our investigations indicate that when the amount of mineral nitrogen in the soil increases, its amount in root-crop increases also (r = 0,51*).

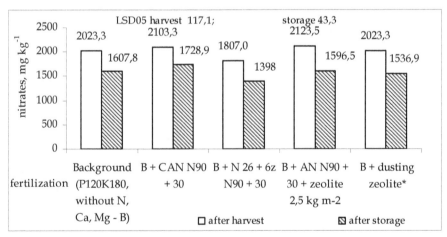

Fig. 7. Influence of fertilizers with prolongation effect on content of nitrates in red beet crop-root. Babtai, 2004–2008.

Field vegetables, gathered later on, as indicate Amr & Hadidi (2001), accumulate fewer nitrates. The influence of nitrogen fertilizers on nitrogen and nitrate amount in root-crops isn't big, but may be important to other properties (Haworth, 1966). Petronienė & Viškelis (2004a) state that the amount of nitrates in red beet increases additionally fertilizing during vegetation, therefore it is important do not delay fertilization. The amount of nitrates can fluctuate from 388–4 880 mg kg^{-1}, fertilizing red beet PK or growing them without fertilizer, and reach even 6 480 mg kg^{-1}, fertilizing them NPK. Influence of fertilizers, in comparison with the influence of meteorological conditions was smaller, and the forms and rates of fertilizers didn't influence nitrate amount (Staugaitis & Dalangauskienė, 2005). In our investigations nitrate amount, when red beet weren't fertilized, was 2023,3 mg kg^{-1} and fertilizing with nitrogen fertilizers didn't influenced the increase of nitrate amount, even though fertilizing with calcium ammonium nitre and ammonium nitre with zeolite their amount was bigger (Fig. 7). The least nitrate amounts both after red beet gathering and after storing were in the root-crops of red beet fertilized with nitrogen fertilizer with zeolite. In red beet root-crops, which were sprinkled with zeolite, nitrate amount was 3,9 % bigger than this in red beet fertilized with nitrogen fertilizer with zeolite, but smaller than after fertilization with nitrogen fertilizer or in red beet grown without nitrogen fertilizer. Polish investigators (Kołota et al., 2007) indicate that fertilizer with nitrification inhibitor (Entec 26) tends to decrease nitrate amounts. According to some investigators (Montemurro et al., 2007), the bigger amount of nitrates in the soil also increases nitrate concentration in plants. Staugaitis (1996) affirms that during storage, because of physiological processes in root-crops, nitrate amount increases, when before storage there are little amount of them (not more than 95 mg kg^{-1}), and decreases when there are more of them. Other investigators indicate that zeolite application in substrate prolongs the time of substrate use and guarantee the bigger and stabile biomass yield and smaller nitrate amount in it (Geodakian, Erofeeva, 1996).

There can be in carrot, as Аутко (2004) indicate, 8–12 % of dry matter, 6–8 % of sugars, 9–12 mg% of carotene, also potassium and microelements – boron and iodine. According to Holden et al. (1999), raw carrot roots contain on average 12 % of dry matter, 4.5 % of sugars, 2.0 % of dietary fiber, 5.7 mg ·100 g^{-1} of β-carotene, 5.9 mg ·100 g^{-1} of vitamin C. According to Ayaz et al. (2007), amounts of dry matter fluctuated in wide limits – from 6,40 % to 11,43 %; nitrates – from 8,1 mg kg^{-1} to 509 mg kg^{-1}. The data obtained in the investigations correspond to the indicated (Fig. 8, 9, 10).

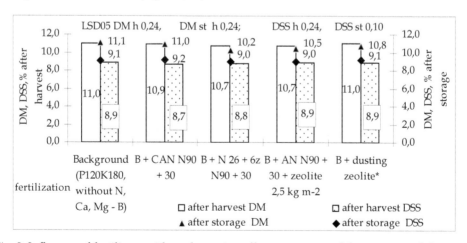

Fig. 8. Influence of fertilizers with prolongation effect on content of dry matter and dry soluble solids in carrot crop-root. Babtai, 2004–2008.

Important carrot quality indices are big amount of sugars and carotenoids, especially β-carotene, and small amount of nitrates (Gajewski et al., 2009). Their amounts depend on growth conditions, genetype and fertilization (Arscott & Tanumihardjo, 2010). During growth period meteorological conditions in our investigations little influenced the changes of biochemical indices in root-crops, with the exception of nitrates. Differentiated soil fertilization with nitrogen doesn't significantly influence the amount of dry matter and dry soluble solids, carotenoids, nitrates and phenols in carrot root-crops (Pokluda, 2006; Pekarskas & Bartaševičienė, 2009; Karklelienė et al., 2007a). The amounts of dry soluble solids can increase when yield gathering is delayed (Suojala, 2000). Data of investigations show that the amounts of dry matter and dry soluble solids fluctuated in very narrow limits and fertilization with nitrogen little influenced their changes (Fig. 8). Both after carrot yield gathering and after storage in root-crops of carrots fertilized with nitrogen their amounts were smaller in comparison with the amounts in the root-crops grown without nitrogen fertilizer. After storage dry matter content in carrot changed in general, depending on kind of storage and variety, and β-carotene content was affected as well. Cold storage showed lower loss (13,57–14,28 %) compared to cellar (20–27,3 %) (Fikselová et al., 2010). Investigation data show that in carrot root-crops fertilized with nitrogen fertilizer with zeolite after storage were was slightly bigger amount of dry soluble solids than there was after yield gathering and fertilization with the investigations nitrogen fertilizers didn't have influence.

Growth and storage conditions, genetype influence sugar (Seljasen, et al., 2011) and carotene (Fikseliová et al., 2010; Gajewski et al., 2007, 2010) amounts in carrot root-crops. When growing carrot cultivar 'Samson', as data of our investigations show (Fig. 9), there were 6,2–6,4 % of sugars, 12,8–13,2 mg% of carotenes and fertilization with the investigated nitrogen fertilizers had little influence on sugar and carotene amounts. After carrot storage sugar and carotene amounts, in comparison with the amounts after yield gathering, little changed. According to polish investigators (Rożek et al., 2000), storing root-crops the amounts of dry sugars and nitrates decrease, but the amounts of phenols increase. Fikselová et al. (2010) indicate that storing in cool place the losses of β-carotene were 13,6–14,3 %, and storing in cellars – 20–27,3 %. Some authors (Belitz et al., 2004) indicate that carrot storage, which doesn't correspond to requirements; increase the disintegration of carotenoids 5–40 %. When carrots are stored at 2 °C and 90 % relational humidity, carotenoid amounts slowly increased during the first 100 days, but later on decreased (Lee, 1986).

Fertilizer rates and the time of their sprinkling, growth conditions influence nitrate accumulation in root-crops (Gajewski et al., 2009). Literature data concerning nitrates accumulation in carrots are differentiated and, according to Pokluda (2006), ranged from 50 to 500 mg kg^{-1} plants grown in Middle Europe region, and Koná (2006) indicates narrower limits (302,5–449 mg kg^{-1}). In the grown experiments in carrot root-crops there were 290,0–308,5 mg kg^{-1} (Fig. 10). Fertilizing with calcium ammonium nitre, the amount of nitrates increased 12,7 mg kg^{-1}, and fertilizing with ammonium nitre with zeolite the increase was insignificant. The least amount of nitrates in root-crops both in autumn, after yield gathering, and after storage, accumulated in carrot root-crops fertilized with nitrogen fertilizer with zeolite. After storage nitrate amounts decreased in root-crops fertilized with calcium ammonium nitre, nitrogen fertilizer with zeolite and grown without nitrogen fertilizer; slight increase, in comparison with the amounts after yield gathering were in carrot root-crops fertilized with ammonium nitre with zeolite. The data of statistical analysis showed

that under our conditions when temperature increased nitrate amount in root-crop increased
also (r = 0,96**), and the increase of precipitation amount during vegetation nitrate amounts
decreased (r = -0,98**). Nitrate amount in root-crops increased, when mineral nitrogen amount
in the soil increased also (r = 0,56*).

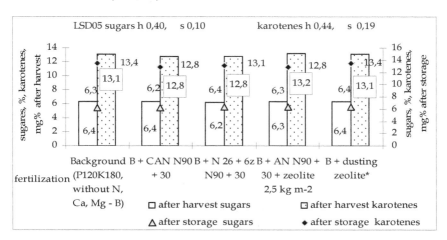

Fig. 9. Influence of fertilizers with prolongation effect on content of sugars and carotenes in
carrot crop-root. Babtai, 2004–2008.

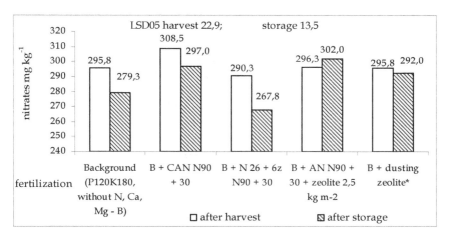

Fig. 10. Influence of fertilizers with prolongation effect on content of nitrates in carrot crop-
root. Babtai, 2004–2008.

Red beet, cabbage, cucumber, potatoes, leaf vegetables accumulate more nitrates. Their
amount during storage little changes. Carrot accumulate little amount of nitrates (Prasad &
Chetty, 2008; Ayaz et al., 2007). Our investigations confirmed this. In red beet root-crops
fertilized with nitrogen fertilizer with zeolite after yield gathering there was found 1807 mg
kg-1, carrot – 290 mg kg-1 of nitrates. After storage nitrates amount of nitrates slightly
decreased: in red beet on the average 442,5 mg kg-1, in carrot – 9,6 mg kg-1.

3.4 Agrochemical properties of the soil

The amount of nutrients in the soil depends on the soil itself, fertilization, and time of investigations (Jalali, 2008; Liet. dirvožemių ..., 1998). Application of natural zeolite, that showed Abdi et al. (2006) increased the available nitrogen, potassium, phosphorus, calcium and magnesium of the medium. Beqiraj (Goga) et al. (2008) maintain that the presence of zeolites ameliorates the physical and chemical quality of soil and by thus can reduce nutrient loss due to leaching by increasing the retention of nutrients and slowly releasing them as needed by soil and plants. The influence of nitrogen fertilization is also determined by many other factors (soil type, texture, redox potential and the content of organic matter, cation exchange capacity, and base saturation ratio, soil content of Ca and Mg as well as heavy metals (Sady & Rożek 2002). Zeolite, which the main component is klinoptilolit, according to Russian investigator (Середина, 2003), decrease soil acidity and this effect is seen in many years. Investigations in Croatia showed that fertilizers with zeolite have little effect upon the reduction of soil acidity by comparison to the applied lime materials, but owing to intensive ion exchange it had a good effect on soil fertility and thereby on the plants yield (Butorac et al., 2002). Soil acidity in red beet crop, when it is established after yield gathering, applying fertilizers little changed (Table 2); in carrot crop (Table 3), after fertilization with nitrogen fertilizer with zeolite, it was slightly smaller in comparison with this when fertilizing with calcium ammonium nitre. Smoleń et al., 2011 showed that nitrogen fertilizers only slightly influenced chemical properties of soil, mainly pH, solubility rate of mineral nutrients in soil environment and, thus, mineral uptake (and accumulation) by carrot plants. The amount of mineral nitrogen in the soil in red beet crop fertilizing with nitrogen fertilizer with zeolite, in comparison with red beet fertilized with calcium ammonium nitre, increased 8.2 %, in comparison with ammonium nitre and zeolite – 3,9 % (Table 2); in carrot crop correspondingly 5,1 % and 5,0 % (Table 3). Investigations carried out in *Endocalcari-Endohypogleyc Cambisol* (*CMg-n-w-can*) showed that fertilizing winter wheat with nitrogen fertilizer with klinoptilolit (N_{120}) at the last leaf stage, in 0–30 cm soil layer there was 37,6 %, and in wax maturity stage – 28,7–32,6 % more mineral nitrogen, in comparison with amount found fertilizing with calcium ammonium nitre. This might because of the reason that plants assimilate less nitrogen, therefore more of it remains in the soil (Mašauskienė & Mašauskas, 2009).

Fertilization/ characteristics	$pH_{(KCl)}$	agile, mg kg^{-1}				Nmin
		P_2O_5	K_2O	Ca	Mg	kg ha^{-1}
Soil agrochemical characteristics before layout of experiments	7,4	371	186	7249	1775	50,4
Background (P120K180, without N, Ca, Mg -B)	7,7	356,5	162,8	7040	1885	42,3
B + CAN N90 + 30	7,8	345,5	168,0	8455	2329	44,1
B + N 26 + 6z N90 + 30	7,8	375	171,8	10260	2834,8	47,7
B + AN N90 + 30 + zeolite 2,5 kg m-2	7,8	380,5	168,0	8795	2379	45,9
LSD05	0,13	49,8	28,8	2337,8	877,8	6,4

Table 2. Influence of fertilizers with prolongation effect on exchange agrochemical properties of soil in red beet crop. Babtai, 2004–2007.

Slightly bigger amount of agile potassium in red beet crop was in the soil fertilized with nitrogen fertilizer with zeolite and in carrot crop – after fertilization with ammonium nitre with zeolite. There was slightly bigger amount of agile phosphorus in both experiments after fertilization with ammonium nitre with zeolite. According to the data of Serbian investigators (Milošević & Milošević, 2010), fertilization with all the rates of investigated fertilizers with zeolite (Agrozel) led to increases in humus, total nitrogen (not significant), potassium and phosphorus (significant to the control – no fertilizer) within all depts. The biggest amount of calcium and magnesium in the soil of red beet crop was fertilizing with nitrogen fertilizer with zeolite, and in carrot crop – fertilizing with ammonium nitre with zeolite. Even though the soil is rich with calcium and magnesium, these elements are antagonists and plats do not assimilate them (Liet. dirvož...., 1998).

Fertilization/ characteristics	$pH_{(KCl)}$	agile, mg kg^{-1}				N min kg ha^{-1}
		P2O5	K2O	Ca	Mg	
Soil agrochemical characteristics before layout of experiments	7,4	371	186	7249	1775	50,4
Background (P120K180, without N, Ca, Mg -B)	7,6	390,3	151,5	7217,5	1822,3	38,8
B + CAN N90 + 30	7,7	407,5	148,2	7960	1912,5	50,6
B + N 26 + 6z N90 + 30	7,8	404,5	156,2	7935,1	2020	53,2
B + AN N90 + 30 + zeolite 2,5 kg m-2	7,8	444,8	166,3	8652,5	2053,8	52,7
LSD05	0,16	30,7	40,17	1349	187,2	15,5

Table 3. Influence of fertilizers with prolongation effect on exchange agrochemical properties of soil in carrot crop. Babtai, 2004–2007.

4. Conclusions

The use of nitrogen fertilizers increased red beet marketable yield 2,1 times, carrot – 14,1 %, when growing without fertilizers there was obtained correspondingly 24,2 t ha^{-1} and 49,0 t ha^{-1}. The biggest marketable red beet (54,7 t ha^{-1}) and carrot (57,6 t ha^{-1}) yield and the output of marketable yield (correspondingly 79,6 % and 71,4 %) were applying nitrogen fertilizer with zeolite (N 26 + 6z $N_{90 + 30}$). Red beet marketable yield, in comparison with this of the red beet grown without nitrogen fertilizer, increased more than twice, carrot – 17,6 %, output of marketable yield correspondingly 21,3 % and 4,1 %.

The results of storage investigation showed that red beet and carrot fertilized with nitrogen fertilizer with zeolite was preserved best of all. After short-time storage there was 48,1 t ha^{-1} of red beet and 50,3 t ha^{-1} of carrot marketable production; storage losses comprised correspondingly 12,0 % and 12,7 %. After long-time storage there was 39,6 t ha^{-1} of red beet and 41,2 t ha^{-1} of carrot marketable production; storage losses comprised correspondingly 27,5 % and 28,4 %. Red beet and carrot sprinkling with zeolite (30 kg t^{-1}) improved their storability. After short-time storage the output of red beet marketable production in comparison with this of the red beet grown without nitrogen fertilizers increased 8,6 %, carrot – 9,0 %; after long-time storage – correspondingly 35,5 % and 29,2 %; i. e., there was obtained additionally 1,7 and 5,1 t ha^{-1}of red beet and 3,8 and 9,5 t ha^{-1}of carrot marketable production.

Dry matter amounts after yield gathering in red beet root-crops fluctuated from 11,9 % to 14,4 %, dry soluble solids – from 11,7 % to 13,5 %; nitrates – from 1807 mg kg^{-1} to 2123,5 mg kg^{-1} and the least their amounts were fertilizing the crop with nitrogen fertilizer with zeolite; sugars – from 7,9 % to 8,4 % and fertilization didn't influenced significantly their amount. In carrot root-crops dry matter amounts fluctuated from 10,7 % to 11,0 %; dry soluble solids – from 8,8 % to 8,9 %; nitrates – from 290 mg kg^{-1} to 308.5 mg kg^1 and the least their amounts were fertilizing the crop with nitrogen fertilizer with zeolite; sugars – from 6,2 % to 6,4 %; carotenes – from 12,8 mg% to 13,2 mg%. After storage in red beet root-crops there was correspondingly 11,8-13,7; 12,0-12,3; 1398-1728,9; 8,6-9,0; in carrot crop – 10,2-11,1; 8,7-8,9; 267,8-302,0; 6,2-6,4 and 12,8-13,4. The least amounts of dry matter, dry soluble solids and nitrates were after fertilization red beet and carrot crop with nitrogen fertilizer with zeolite. Fertilization with the investigated nitrogen fertilizers didn't influence significantly the amounts of sugars and carotenes.

Fertilizing with nitrogen fertilizer with zeolite soil acidity decreased, and the amounts of mineral nitrogen were bigger. The amounts of agile potassium, phosphorus, calcium and magnesium in the soil fertilizing with nitrogen fertilizer with zeolite were bigger than fertilizing with other investigated fertilizers.

Zeolite is the suitable means for soil properties and yield quality improvement.

5. Acknowledgements

Authors are grateful to Lithuanian State Foundation of Science and Studies, SC "Achema" and JSC "Elega" for financial support carrying out investigations.

6. References

Abdi Gh., Khosh-Khui M, Eshghi S. (2006). Effects of natural zeolite on growth and flowering of stawberry (*Fragariaxananassa* Duch.). *International Journal of Agricultural Research*, vol. 1, № 4 (August, 2006), pp. 384-389, ISSN 1816-4897

Alexandrescu A., Gawriluta I., Buzdugan C., Beldiman G., Borlan Z. (2000) Agrochemical possibilities of diminishing nitrate content in edible parts of plants. *Life sciences and geosciences*, vol. 2, № 1 (March, 2000), pp. 77-80, ISSN: 1454-8267.

Ambruster T. (2001). Clinoptilotite-heulandite: applications and basic research, *Zeolites and Mesoporous Materials at the dawn of the 21st century, Proceedings of the 13 International Zeolite Conference*, pp. 13-27, ISBN: 978-0-444-50238-4, Montpellier, France, Eds. Galarneau A., Fajula F., Di Renzo F., Vedrine J.

Amr A., Hadidi N. (2001). Effect of cultivar and harvest date on nitrate (NO$_3$) and nitrite (NO$_2$) content of selected vegetables grown under open field and greenhouse conditions in Jordan. *Journal of Food Composition and Analysis*, vol. 14, № 1 (February, 2001) , pp. 59-67(9), ISSN ISSN 0889-1575

Ancuta A., Mažeika R., Staneika E., Buzienė A., Jančiauskas M., Mašauskas V., Venckūnas V., Balčiuvienė E., Arlauskas R. (2011). Amonio trąšos su karbonatais ir ceolitu. Patentas LT 5723 B. Patento paskelbimo data 2011-04-26

AOAC. 1990a. Solids (soluble) in fruits and fruit products. In: *Official Methods of Analysis*. 15th edition, Helrich K. (ed.). AOAC Inc., Arlington, VA: 915., ISBN 0-935584-42-0

AOAC. 1990b. Sucrose in fruits and fruit products. In: *Official Methods of Analysis*. 15th edition. Helrich K (ed.), AOAC Inc., Arlington, VA: 922. ISBN 0-935584-42-0

Arscott S. A., Tanumihardjo S. A. (2010). Carrots of many colors provide basic nutrition and bioavailable phytochemicals acting as a functional food. *Comprehesive reviews in food science and food safety,* vol. 9, № 2 (March, 2010), pp. 223-239, ISSN 1541-4337 (online)

Ayaz A, Topçy A., Yurttagul M. (2007). Survey of Nitrate and Nitrite Levels of Fresh Vegetables in Turkey. *Journal of Food technology,* vol. 5, № 2 (March, 2007), pp. 177-179, ISSN 1648-8462 (print), ISSN 1993-6036 (online)

Babik I., Elkner K. (2002). The effect of nitrogen fertilization and irrigation on yield and quality of broccoli. *Acta Horticulturae,* vol. 571, № (February, 2002), pp. 33-43, ISSN 0567-7572

Badelek E., Adamicki F., Elkner K. (2002). The effect of temperature, cultivar and root size on quality and storage ability of red beet. *Vegetable crops research bulletin,* 2002, vol. 56, № (, 2002), pp. 67–76, ISSN 1506-9427

Beqiraj (Goga) E., Fran Gjoka F.,, Fabrice Muller F., Baillif P. (2008). Use of zeolitic material from Munella region (Albania) as fertilizer in the sandy soils of Divjaka region (Albania). *Carpthhian Journal of Earth and Environmental Sciences,* vol. 3, №. 2 (October, 2008), pp. 33 – 47, ISSN 1842 – 4090 (print), ISSN: 1844 - 489X (online)

Belitz H. D., Grosch W., Schieberle P. (2004). Vitamins. In: *Food Chemistry*. 3rd revised., ed. Belitz H.D., Grosch W., Schieberle P., pp. 409-426, Springer-Verlag, ISBN 3-540-40817-7, Berlin, Heidelberg, New York,.

Butkuvienė E., Kuodienė R. Daugėlienė N. (2006). Effect of wood ash rates on red beet and potato yield and their quality. *Sodininkystė ir daržininkystė,* vol 25, №1 (March, 2006), pp. 207-215, ISSN 0236-4212

Butorac A., Filipan T., Bašić F., Butorac J., Mesić M., Kisić I. (2002). Crop response to the application natural amendments based on zeolite tuff. *Rostlinnà Výroba,* vol. 48, № 3 (March, 2002)), pp.118-124, ISSN 0370-663X

Challinor P. F., Le Pivert J. M., Fuller M. P. (1995). The production of standard carnations of nutrient loaded zeolite. *Acta Horticulturae,* vol. 401, №1 (October, 1995), pp. 293-300, ISSN 0567-7572.

Conry M. J. (1997). Effect of fertilizer N on the grain yield and quality of spring barley grown on five contrasting soils in Ireland. *Biology and Environment,* vol. 97B, № 3 (December, 1997), pp. 185-196, ISSN 0791-7945

Davies B. H. 1976. Carotenoids. In: *Chemistry and Biochemistry of Plant Pigments*, vol. 1, ed. Goodwin T. W., pp. 35-165, Academic Press, ISBN 10 0122899016, London, New York

Fikselová M., Mareĉek J, Mellen M. (2010). Carotenes content in carrot roots (*Daucus carota* L.) as affected by cultivation and storage. *Vegetable Crops Research Bulletin,* vol. 73, № , (January, 2011), pp. 47-54, ISSN 1506-9427 (print), ISSN 1898-7761 (online)

Manuals of Food Quality Control: Food Analysis, General Techniques, Additives, Contaminants and Composition (1986), by Food and Agriculture Organization of the United Nations, ISBN 9251023999, Hardcover, FAO

Fritz, D., Weichmann, J. (1979). Infuence of the harvesting date of carrots on quality and quality preservation. *Acta Horticulturae,* vol. 93, № 1 (December, 1979), pp. 91-100, ISSN 0567-7572

Fustec J., Lesuffleur F., Mahieu S., Cliquet J. B. (2010). Nitrogen rhizodeposition of legumes. A review. *Agronomy for. Sustainanable Development,* vol. 30, № 1 (January-March, 2010), pp. 57-66, ISSN 1774-0746, eISSN 1773-0155

Gajewski M., Węglarz Z., Sereda A., Bajer M., Kuczkowska A., Majewski M. (2009). Quality of carrots grown for processingas affected by nitrogen fertilizationand harvest term. *Vegetable Crops Research Bulletin,* vol. 70, № (July, 2009), pp. 135-144, ISSN 1506-9427

Gajewski M., Węglarz Z., Sereda A., Bajer M., Kuczkowska A., Majewski M. (2010) Carotenoid Accumulation by Carrot Storage Roots in Relation to Nitrogen Fertilization Level. *Notulae Botanicae Horti Agrobotanici Cluj Napoca,* vol. 38, № 1 (January, 2010), pp. *71-75,* ISSN 0255-965X (print); 1842-4309 online)

Gajewski M., Szymczak P., Elkner K., Aleksandra Dąbrowska A., Kret A., Honorata Danilcenko H. (2007). Some aspects of nutritive and biological valueof carrot cultivars with orange, yellowand purple-coloured roots. *Vegetable Crops Research Bulletin,* vol. 67, № (December, 2007), pp. 149-161, ISSN 1506-9427

Geodakian R. O., Erofeeva T.V. (1996). The effectiveness of using biohumus for growing plants under autonomous conditions. *Aviakosmicheskaia i ekologicheskaia meditsina,* vol. 30, № 3 (June, 1996), pp 39-43, ISSN 0233-528X (print)

Grattan S. R., Grieve C. M. (1998). Salinity-mineral nutrient relations in horticultural crops. *Scientia Horticulturae,* 1998, vol. 78, №1-4 (November, 1998), pp.127–157, ISSN 0304-4238.

Gül A., Eroğul D., Ongun A. R. (2005). Comparison of the use zeolite and perlite as substrate for crisp-head lettuce. *Scientia Horticulturae,* vol. 106, № (November, 2005), pp. 464-471, ISSN 0304-4238

Haworth F., Cleaver T.J., Bray J.M. (1966). The effects of different manurial treatments on the yield and mineral composition of red beet. *The Journal of Horticultural Science & Biotechnology,* vol. 41, №3 (July, 1996), pp. 243-256, ISSN 1462-0316

Holden, J. M., Eldridge, A. L., Beecher, G. R., Buzzard, I. M. Bhagwat, S. A., Davis, C. S., Douglass, Larry W., Gebhardt, S. E., Haytowitz, D. B., Schakel, S. 1999. Carotenoid content of U.S. coods: An update of the database. *Journal of Food Composition and Analysis,* 1999, vol 12, № 3 (September, 1999), pp.169-196, ISSN: 0889-1575

Ilsildar A.A. (1999). Effect of the addition of zeolite to the soil on nutrification. *Turkish Journal of Agriculture and Forestry,* vol. 23, № 3 (Jule, 1999), pp. 363-368, ISSN 1300-011x.

Jalali M. (2008). Nitrate concentrations in some vegetables and soils in Hamadan, western Iran. *Archives of Agronomy and Soil Science,* vol. 54, № 5 (August, 2008), pp. 569-583, ISSN 0365-0340

Yolcu H., Seker H., Gullap M. K., Lithourgidis A., Gunes A. (2011). Application of cattle manure, zeolite and leonardite improves hay yield and quality of annual ryegrass (*Lolium multiflorum* Lam.) under semiarid conditions. *Austarlian Journal of Crop Science,* vol 5, № 8 (, 2011), pp. 926-931, ISSN 1835-2707.

Karklielienė R., Duchovskienė L.,Dambrauskienė E., Bobinas Č. (2007). Investigation of productivity of seed stalks of edible carrot and red beet Lithuanian cultivars. *Sodininkystė ir daržininkystė,* vol. 26, № 3 (August, 2007), pp. 198-207, ISSN 0236-412

Karklelienė R., Juškevičienė D., Viškelis P. (2007a). Productivity and quality of carrotr *(Daucus sativus* Röhl.) and onion *(Allium cepa* L.) cultivars and hybrids. *Sodininkystė ir daržininkystė,* vol 26, № 3 (August, 2008), pp. 208-216, ISSN0236-4212

Karklelienė R., Viškelis P., Radzevičius A., Duchovskienė L. (2009). Evaluation of productivity and biochemical composition of perspective red beet breeding number. *Acta Horticulturae,* vol 830, № 1 (June, 2009), pp.255-260, ISSN 0567-7572

Kołota E., Adamczewska-Sowinska K., Kręzel J. (2007). Suitability of Entec 26 as source of nitrogen for red beet and celeriac. *Vegetable Crops Research Bulletin,* 2007, vol. 67, № (December), pp. 47-54, ISSN 1506-9427 (print), 1898-7761 (Online)

Kóňa, J. (2006). Nitrate accumulation in different parts of carrot root during vegetation period. *Acta Horticulturae et Regiotecturae,* vol. 9, №1 (, 2006), pp. 22-24, ISSN 1335-2563

Klotz K. J., Fernando L. Finger F .J. (2004). Impact of temperature, length of storage and postharvest disease on sucrose catabolism in sugarbeet. *Postharvest Biology and Technology,* vol. 34, № 1, (October, 2004), pp. 1-9, ISSN 0925-5214

Lairon D. (2010). Nutritional quality and safety of organic food. A review. *Agronomy for Sustainable Development,* vol. 30, № 1 (January-March, 2010), pp. 33-41, ISSN 1774-0746, eISSN 1773-0155

Lee C. I. (1986) Changes in carotinoid content of carrots during growth and post-harvest storage. *Food chemistry,* vol. 20, № 4 (October, 1986), pp. 285-293, ISSN 0308-8146

Li Z., Alessi D., Allen L. (2002). Influence of quaternary ammonium on sorption of selected metal cations onto cilnoptilolite zeolite. *Journal of Environmental Quality,* vol. 31, № (July-August, 2002), pp. 1106-1114, ISSN 0047-2425

Li Z. (2003). Use of surfactant-modified zeolite as fertilizer carriers to control nitrate release. *Microsporous and Mesoporous Materials,* vol. 61, № 1 (July, 2003), pp. 181-188. ISSN 1387-1811

Lietuvos dirvožemių agrocheminės savybės ir jų kaita. (1998). Sudaryt. J. Mažvila., Lietuvos žemdirbystės institutas, ISBN 9986-527-47-3, Kaunas

Lobova B. P. (2000). Use of mineral raw material containing zeolite in agriculture. *Agrokhimiya,* 2000, vol. 6, № (, 2000), pp 78-91, ISSN 0002-1881

Malekian R., Abedi-Koupai J., Eslamian S. S. (2011). Influences of clinoptilolite and surfactant-modified clinoptilolite zeolite on nitrate leaching and plant growth. *Journal of Hazardous Materials,* vol. 185, № 2-3, (January, 2011), pp. 970-976, ISSN 0304-3894

Malnou C. S., Jaggard K.W., Sparkes D.L. (2008). Nitrogen fertilizer and the efficiency of the sugar beet crop in late summer. *European Journal of Agronomy,* vol. 28, № 1 (January, 2008), pp. 47-56, ISSN 1161-0301

Mašauskienė A., Mašauskas V. 2009. Azoto trąšų su klinoptilolitu įtaka mineralinio azoto išplovimui iš vandens. *Žemdirbyste-Agriculture,* vol. 96 № 4 (December, 2009), pp. 32-36, ISSN 1392-3196

Mažeika R., Cigienė A., Tatariškinaitė L. 2008. Nitratinių trąšų tobulinimas ir naujų sukūrimas. In *Nitratinių trąšų tobulinimas, naujų sukūrimas ir jų efektyvumo įvertinimas,* R. Mažeika and V. Mašauskas (Eds.), p. 18-29, AB „Achema", ISBN 978-9986-9127-1-1, Jonava Metodiniai nurodymai nitratams nustatyti augalininkystės produkcijoje. (1990). Vilnius.

Montemurro F., Maiorana M., Lacertosa G. (2007). Plant and soil nitrogen indicators and performance of tomato grown at different nitrogen fertilization levels. *Journal of Food, Agriculture and Environment*, vol. 5, № 2 (April, 2007), pp. 143–148, ISSN 1459-0255.

Milosević T., Milosević N. (2010). The effect of organic fertilizer, composite NPK and clinoptilolite on changes in the chemical composition of degraded Vertisol in Western Serbia. *Carpathian Journal of Earth and Environmental Sciences*, vol. 5, № 1, (April, 2010), pp. 25 – 32, ISSN 1842-4090

Pekarskas J., Bartaševičienė D. (2009). Ekologiškai augintų morkų veislių derlingumas ir biocheminė sudėtis. *Sodininkystė ir daržininkystė*, vol 28, № 4 (June, 2009), pp. 99-105, ISSN 0236-4212

Petronienė O. D., Viškelis P. (2004). Biochemical composition and preservation of various red beet cultivars. *Sodininkystė ir daržininkystė*, vol. 23, № 3, (September, 2004), pp. 89–97, ISSN 0236-4212.

Petronienė O. D., Viškelis P. (2004a). Įvairių veislių tipų ir grupių raudonųjų burokėlių (*Beta vulgaris* L.) biocheminė sudėtis. *Maisto chemija ir technologija*, vol. 38 № (,2004), pp. 42-47, ISSN 1392-0227

Pokluda R. (2006). An assessment of the nutritional value of vegetables using an ascorbate-nitrate index. *Vegetable Crops Research Bulletin*, vol. 64, № (, 2006), pp.: 28-37, ISSN 1506-9427

Polat E., Karama M., Demir H., Onus N. (2004). Use of naturale zeolite (clinoptilolite) in agricultural. *Journal of Fruit and Ornamental Plant Research*, vol 12, № special ed., (, 2004), pp. 183-189, ISSN 1231-0948

Porro D., Dorigatti C., Ramponi M. (2002). Can foliar application modify nutritional status and improve fruit quality results on apple in northeastern Italy. *Acta Horticulturae*, vol 594, № 1 (November , 2002), pp. 521–526, ISSN 0567-7572

Prasad S., Chetty A. A. (2008). Nitrate-N determination in leafy vegetables: Study of the effects of cooking and freezing. *Food Chemistry*, vol. 106, №2 (January, 2008), pp. 72-780, ISNN 0308-8146

Raju P. S., Chauhan O. P., Bawa A S. (2010). Postharvest handling systems and storage of vegetables. *In Handbook of vegetables preservation and processing*. Sinha N. K. , Hui Y H., Özgül E., Siddinq M., Ahmed J. (eds), ISBN 978-0-8138-1541-1, Willey Black, Singapūre

Ramesh K., Dendi Damodar Reddy D. D., Biswas A. K., Rao A. S. (2011). Chapter 4 - Zeolites and their potential uses in agriculture. *Advances in Agronomy*, 2011 vol. 113, № (September, 2011), pp. 215-236, ISSN : 0065-2113

Rehakovà M., Čuvanovà M., Dzivàk M., Rimàr J., Gaval'ovà Z. (2004). Agricultural and agrochemical uses of natural zeolite of the clonoptilolite type. *Current Opinion in Solid State and Materials Science*, vol. 8, № 6 (December, 2004), pp. 397-404, ISSN 1359-0286

Rines R. H., Toth L., Rines-Toth S. 2006. Method of and products for promoting improved growth of plants and more water-efficient growing soil or other media and the like with zeolite crystals treated with prefarably water-based plant derived nutrient extractions and the like. *US Patent 7056865. US Patent Issued of June 6, 2006*

Rożek S., Leja M., Wojciechovska R. (2000). Effect of differentiated nitrogen fertilization on chantes of certain compounds in stored carrot roots. *Folia Horticulturae*, vol. 12, № 2 (March, 2000), pp. 21-34, ISSN 0867-1761.

Ryakhovskaja N. J., Gainatulina V. V. (2009_. Potato and oat yield in short-cycle crop rotation with zeolite application. *Russian Agricultural Sciences*, vol. 35, № 3 (September, 2009), pp. 153-155, ISSN 1068-3674

Rubatzky V. E., Quiros C. F., Simon P. W. (1999). *Carrots and related vegetable umbelliferae.* Publisher: C A B Intl, University of California, Davis, University of Viskonsin, Madison, pp. 245–256, ISBN 0851991297/0-85199-129-7, USA

Sady W., Rożek S. (2002). The effect of physical and chemical soil properties on the accumulation of cadmium in carrot. *Acta Horticulturae*, vol. 571 № 1 (February, 2002), pp. 72-75, ISSN 0567-7572

Samartzidis C., Awada T., Maloupa E., Radoglou K., ConstantinidouH. I. A. (2005). Rose productivity and physiological response to different substrates for soil-less culture. *Scientia Horticultura*, vol. 106, № 2 (September, 2205), pp. 203 – 212, ISSN 0304-423.

Sakalauskas A., Zalatorius V., Malinauskaitė S. (2005). Biometrinių matmenų įtaka morkų prekiniam paruošimui. *Sodininkystė ir daržininkystė*, vol. 24, № 1 (April, 2005), pp. 72- 79, ISSN 0236-4212

Santamaria P. (2006). Nitrate in vegetables: toxicity, content, intake and EC regulation. *Journal of the Science of Food and Agriculture*, vol 86, № 1 (January, 2006), pp. 10-17, ISSN 0022-5142

Saure M.C., 2005. Calcium translocation to fleshy fruit: its mechanism and endogenous control. *Scientia Horticulturae*, vol. 105, № 1 (May, 2005), pp. 65-89, ISNN 0304-4238

Scholberg J., McNeal B.-L., Boote K.-J., Jones J. W., Locascio S. J., Olson S. M. (2000). Nitrogen stress on growth and nitrogen accumulation by field-grown tomato. *Agronomy Journal*, 2000, vol. 92, № 1 (January-February, 2000), pp. 159-167, ISSN 0002-1962

Seljåsen R., Bengtsson G., Hoftun H., Vogt G. 2001. Sensory and chemical changes in five varieties of carrot in response to mechanical stress and postharvest. *Journal of the Science of Food and Agriculture*, vol. 81, № 4 (March, 2001), pp. 436-447, ISSN 0022-5142

Smoleń S., Sady W., Wierzbińska J. (2011). The effect of various nitrogen fertilization regimes on the concentration of thirty three elements in carrot (*Daucus carota* L.) storage roots. *Vegetable Crops Research Bulletin*, vol. 74, № (August, 2011), pp. 61-76, ISSN 1506-9427(print), 1898-7761 (online)

Sorensen J. (1999). Nitrogen effect on vegetable crop production and chemical composition. *Acta Horticulturae*, vol. 506, № 1 (December, 1999), pp. 41-50, ISSN 0564-7572

Staugaitis G.(1996). Burokėlių tręšimu azotu sistema. *Sodininkystė ir daržininkystė*, vol. 15, № (July, 1996), pp.63-75, ISNN 0236-4212

Staugaitis G., Dalangauskienė A. (2005). Vienanarių ir kompleksinių trąšų įtaka raudoniesiems burokėliams. *Sodininkystė ir daržininkystė*, vol. 24, № 2 (July, 2005), pp.133-144, ISNN 0236-4212.

Staugaitis G., Tarvydienė A. (2004). Sėklų ir trąšų normų įtaka įvairių veislių tipų raudonųjų burokėlių derliui. *Sodininkystė ir daržininkystė*, vol. 23, № 1 (April, 2004), pp. 144-152, ISSN 0236-4212.

Suojala T. 2000. Variation in sugar content and composition of carrot storage roots at harvest and during storage. *Scientia Horticulturae*, vol. 85, № 1, (July, 2000), pp. 1–19. ISNN 0304-4238

Szura A., Kovalska J., Sady W. (2008). Biological value of red beet in relation to nitrogen fertilization. *Vegetable Crop Research Bulletin,*, vol. 68, № (,2008), pp. 145-153, ISSN 1506-9427(print), 1898-7761 (online)

Tarvydienė A., Petronienė D. (2003). Raudonųjų burokėlių derlius ir biocheminė sudėtis Rytų, Vidurio ir Vakarų Lietuvos agroklimatinėse zonose. *Sodininkystė ir daržininkystė*, vol. 22, № 1 (March, 2003), pp 108-120, ISSN 0236-4212

Tei F., Benincasa P., Guiducci M. (2002). Effect of N availability on growth, N uptake, light interception and pfotosyntetic activity in processing tomato. *Acta Horticulturae*, vol. 571, № 1 (February, 2002), pp. 209-216, ISSN 0567-7572.

Tucker W. G., Ward C. M., Davies A. C.(1977). W. An assessment of the long term storage methods for beetroot. *Acta Horticulturae*, vol. 62, №.1 (June, 1977), pp. 169-180, ISSN 0567-7572.

Ugrinovic K. (1999). Effect of nitrogen fertilization on quality and yield of red beet (*Beta vulgaris var. Conditiva* Alef.). *Acta Horticulturae*, vol. 506, № 1 (February, 2002), pp. 99-104, ISSN 0567-7572.

Uher A. (2004). Use of zeolites for recultivation of sandy soils in horticulture. *Acta Horticulturae et Regiotecturae*, vol. 7 № 1 (May, 2004), pp. 32-36 ISSN 1335-2563

Zalatorius V., Viškelis P. (2005). Papildomo juostinio tręšimo per lapus įtaka morkų produktyvumui ir laikymuisi. *Sodininkystė ir daržininkystė*, vol. 24, № 1 (March, 2005). Pp. 72-79, ISSN 0236-4212

Zdravkovic M., Damjanovic M., Corokalo D. (1997). The influence of fertilization on yield of different carrot varietes. *Acta Horticulturae*, vol. 462, № 1 (December, 1997), pp. 93–96, ISSN 0567-7572.

Wang Z. H., Li S. H, Malhi S. (2008). Effects of fertilization and other agronomic measures on nutritional quality of crops. Review. *Journal of the Science of Food and Agriculture*, vol. 88, № 1 (October, 2008), pp. 7-23, ISSN 0022-5142

White P. J., Martin R. Broadley M. R.(2003). Calcium in Plants. *Annals of Botany*, vol. 92, № 4 (October, 2003), pp: 487–511. ISNN 0305-7364

Workneh T. S., Ostoff G., Steyn M. S. (2011). Physiological and chemical quality of carrots subjected to prie- and postharvest treatments. *African Journal of Agricultural Research*, vol. 6, № 12 (June, 2011), pp. 2715-2724, ISSN 1991-637x

Алексеева Т. П., Перфильева В. Д., Криницын Г.Г. (1999). Комплексное органно-минеральное удобрение пролонгированного действия на основе торфа. *Химия растительного сырья*. № 4 (February, 1999), pp. 53- 59, ISSN, 1029-5151 (print), 1029-5143 (online)

Аутко А.А. (2004). *В мире овощей*. УП «Технопринт», ISBN 985-464-635-1, Минск

Дерюгин И. П., Кулюкин А. Н. (1988). *Агрохимические основы системы удобрения овощных и плодовых культур*, Агропромиздат, ISBN 5-10-000402-9, Москва.

Середина В.П. (2003). Агроэкологические аспекты использования цеолитов как почвоулучшителей сорбционного типа и источника калия для растений. *Известия Томского политехнического университета*, 2003, vol. 306. № 3 (mėnuo), pp. 56-60, ISSN

2

Effect of the Climate and Soil Characteristics on the Nitrogen Balance in the North of Algeria

N. Bettahar

Laboratory Water & Environement, Department of Hydraulic,
University Hassiba Ben Bouali, Chlef,
Algeria

1. Introduction

The regular growth of the nitrate concentrations observed in the superficial and underground waters since the years 70 is a topic of preoccupation (Gomez, 2002). This general increase is largely imputed to the agricultural activities, which knew deep modifications.

The agricultural pollution is problematic because of its diffuse character. Of this fact, the solutions can be only preventive while reconciling effective agriculture (establishment of balance to assure the good management of nitrogen in soil) and quality of water using regimentations.

Numerous research elaborated methods of balances of the nourishing elements to develop a lasting agriculture (Parris, 1998). However, the principles that found these balances are very variable, so much to the level of the sought-after precision (choice of the fluxes and parameters took in account and simplifying hypotheses), of the scales of the study (country, exploitation, rotation, parcel), or the length of observation (season, year, etc.) (Van Bol, 2000).

The nitrogenous balance method permits the nitrogenous excess calculation that constitutes the quantity of available remaining nitrogen in soil, capable to be leached toward the aquifer. We have three predominant types of nitrogen contribution: the contributions bound to the mineral fertilizers and the irrigation (Benoît et al., 1997; Sivertun & Prange, 2003; Delgado & Shaffer, 2002), the contributions bound to the breeding and finally the contributions bound to the municipal waste water.

However, the fate of nitrogen in the middle depends on the type of soil, of the type of culture, of the bacterial activity in soil, of the out-flow of water in the matrix of soil and the environmental conditions (Pinheiro, 1995). There are strong interactions between these factors, but environmental conditions, as the temperature, the humidity, the pH, the dissolved oxygen, will play essential roles.

Large quantities of inorganic and organic N- fertilizers are applied each year in agricultural areas (Feng et al., 2005; Elmi et al., 2004; Sivertun et Prange, 2004; Delgado et Shaffer, 2002), which increases the threat from NO_3^- contamination in groundwater. Several processes can manage these quantities of nitrogen in the nature (Tremblay et al., 2001).

The losses of nitrogen are bound to the absorption by the culture that depends on the climate, of the nature of the cultures (Weier, 1992), their stage of growth (Haynes, 1986), the content of the other nourishing elements in soil and the availability of the soil water (Tremblay et al., 2001).

Machet et al., (1987) note that 40 to 60% of nitrogen absorbed by the plants come from the soil nitrogen. Tremblay et al., (2001) note also that even in the best conditions, the plants are not capable to absorb more than 80% of nitrogen contained in the fertilizer applied. The availability of water in soil encourages the absorption of water and nitrates by the plants, various process, either the volatilization, the denitrification and the leaching don't let the rest accessible.

The losses by volatilization depend on conditions of soil (pH, capacity of exchange, porosity, humidity) and of the climatic conditions (Sommer et al., 1991). At the time of mineral fertilizer's application, the losses are less important. The fertilizers which contain the urea entail the volatilization as well as the nitrate of ammonium, the diammonic sulphate and the chloride of ammonium. In this case, the losses can reach 40 to 50% of nitrogen applied in the conditions of chalky soil, of pH> 7.5 and of elevated temperature (Tremblay et al., 2001; Hargrove, 1988).

However, the urea remains the fertilizer that frees the strongest quantities of ammonia in the atmosphere, producing 72% of the quantities freed by the application of fertilizers (Environment Canada, 2000).

The denitrification occurs in soils poor in oxygen, as the swamps, the peaty soils and soils badly drained and is encouraged by elevated temperatures (> 15°C). It is inhibited for a temperature lower to 8°C, what explains the existence of the maximal nitrogen stock in winter (Payraudeau, 2002).

The denitrification is influenced by the environmental conditions as the temperature, the humidity, the content in organic matter of soil (Rassam et al., 2008; Addy et al., 1999), the availability in oxygen (Smith & Tiedje, 1979), the morphology of soil, the pH (Standford et al., 1975) or the activity of the microorganisms (Firestone, 1982). This process can be increased strongly in irrigated cultures that permit to gather several favorable conditions: the presence of fertilizers, the elevated humidity level, the organic product contribution at periods where the temperature is favorable to the microbial activity.

In Western middle Cheliff valley (North of Algeria), Agriculture is the dominant activity. The agricultural land surface constitutes 67% of the total and the main cultures are arboriculture and the garden farming. The alluvial aquifer situated in centre of this zone is exploited for the drinking water supply, the irrigation and industry. In this study, we try at first to show the spatial evolution of nitrate, through a map established by ordinary kriging method for the year 2004 in periods of high water.

Secondly, we try to estimate, for this year, the total contribution of nitrogen present on soils of the valley. It supposes to estimate nitrogen brought by N-fertilizers used extensively in garden farming, potatoes in particular, by water of irrigation from individual wells, by breeding and by municipal waste water.

2. Materials and methods

2.1 Characteristics of study area

The zone of study is located in North-Western Algeria, approximately 200 km to the west of Algiers, and 30 km away from the Mediterranean. It occupies a territory of 300 km² approximately in the basin of Western Middle-Cheliff (Fig. 1).

The area is characterized by a semi-arid climate. The infiltration deduced from the surplus water constitutes 7% (25 mm) of total rainfall (361 mm).

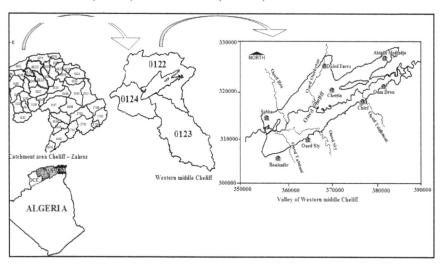

Fig. 1. Location map of study area

The alluvial aquifer situated in centre of this zone is formed by coarse alluvia of age Pliocene Quaternary forming the embankment of the valley (Fig. 2).

It is exploited for the drinking water supply, the irrigation and industry with an annual volume of 15 Million m³. The depth of water varies between 4 and 65 m with an average oscillating around 22 m.

2.2 Types of soil

Two big wholes of soils are observed:

-   Soils of the borders of the valley; They have a balanced texture (25% sand, 35% silt and 40% clay), are deep and structured and present high hydraulic conductivity with elevated pH (8) (Scet Agri, 1984).
-   Soils of plain, alluvial, with variable texture, locally clayey. The heavy soils (> 40% of clay on average) are important on the more recent alluvial formations as the plain of Boukadir, northwest of Wadi Sly and southwest of Ech-Chettia. These soils are chalky (21% of $CaCO_3$) with a very high pH (8.3).

The C/N report for the two types of soil denotes a good mineralization, of a weak rate of nitrogen mineralizable bound to the weak content in organic matter.

The agricultural land surface constitutes 67% of the total of which 65%, either 11700 ha, are irrigated effectively. The main cultures are arboriculture and the garden farming; this last, located near the borders of area study, is a large consumer of N- fertilizers and irrigation relies mainly on groundwater.

Fig. 2. Geological context of study area (Perrodon (1957) & Mattauer (1958))

3. Results and discussion

3.1 Space-time evolution of nitrates

3.1.1 Origin of the data

The study of the evolution of the contents nitrates was undertaken to highlight the former stages of enrichment of water of the studied aquifer which has ends in the current situation. We collected near the service of the National Agency of hydraulic resources the chemical analyses of the major elements corresponding to the taking away carried out on collecting belonging to the inspection network managed by this organism. The data are available for the years 1992, 1993, 1994, 1997, 1998, 1999, 2002, 2003, in addition to the results of analyses which we carried out to us even during the year 2004 in periods of high and low water.

3.1.2 Evolution of the nitrate concentrations groundwater between 1992 and 2004

Four classes of nitrate concentrations are distinguished for the campaigns previously described (Fig. 3):

- Lower than 25 mg/l (represented in blue): water of optimal quality to be consumed;
- Between 25 and 50 mg/l (represented in green): acceptable water of quality to be consumed;
- Between 50 and 100 mg/l (represented in orange): non-drinking water, disadvised for nourrissons and women enclosure, a treatment of potabilisation is necessary before distribution;

- Higher than 100 mg/l (represented in red): water disadvised for all the categories of population, the potabilisation is impossible.

It clearly appears, according to the figure 3, that at the beginning of the years nineteen, more half and until two thirds of the sampled wells offered water of optimal quality to acceptable for drinking. On the contrary, the percentage of well with which water is excessively charged of nitrates represent, with the average, just 9% of the whole of these wells. At the end of this decade, the percentages of well pertaining to the first two classes narrowed with the profit of the third classifies in particular (which represented more than 34% to the average) and in a less way of the last (12% approximately).

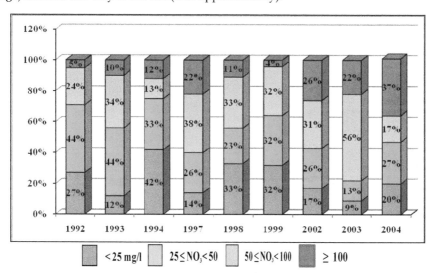

Fig. 3. Proportion of well per class of nitrate concentration

The contracting of the classes of good quality for drinking is accentuated more during the years 2000 to reveal clearly the class of contents nitrates higher than 100 mg/l with a percentage of rather significant well (around 28%).

The number of offering well of non-drinking waters (> 50 mg/l) is thus significant and rises to 63% of the whole of the sampled wells.

It is clear according to this report that the total tendency of the evolution of the nitrate concentrations of water of this aquifer represents a progressive temporal degradation of the quality of this water intended for drinking and/or the irrigation.

The description of the current state of the water quality of this aquifer proves also significant. This is why, a space distribution of the maximum contents nitrates is established for the year of study (between high and low waters of the year 2004).

3.1.3 Maximum contents nitrates of the year 2004

The maximum contents nitrates of the 34 wells to both campaigns of the year 2004 (high and low waters) are distributed in the following way (Fig. 4):

- The number of wells whose maximum content is higher than 25 mg/l is 28, that is to say 82%,
- The number of points whose maximum content is higher than 50 mg/l is of 19, that is to say 56%,
- The number of wells whose maximum content is higher than 100 mg/l is 11, that is to say 32%

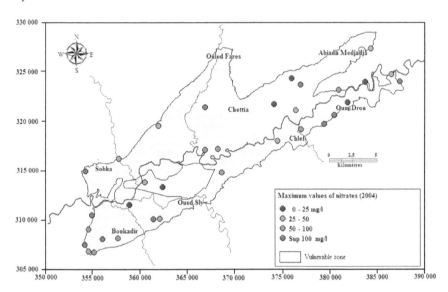

Fig. 4. Space distribution of the maximum contents nitrates (year 2004)

3.1.4 Evolution analysis of the contents nitrates from 1992 to 2004

The number of common wells whose respective contents are indicated is 20. From 1992 to 2004:

Variation of NO₃ (mg/l)	A number of points whose evolution of the contents (Δ) is				
	in reduction		stable	in increase	
	Δ≤-5	-5<Δ<-1	-1≤Δ≤+1	1<Δ<+5	+5≤Δ
Number of wells	7	1	0	0	12
	8		0	12	

Table 1.

- the number of wells whose nitrate content is higher or equal to 50 mg/l evolved from 30% to 45%.
- the evolutions of nitrate contents for the same 20 wells are distributed in the following way:

This highlights:

- A tendency to degradation on 12 wells (60 %) with increase of content higher than 1mg/l (Fig. 5),
- A tendency to the improvement on 8 wells (40 %) with a reduction of content least 1mg/l.

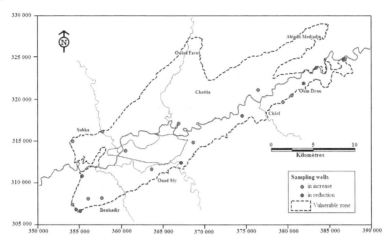

Fig. 5. Evolution of the contents from 1992 to 2004

The annual evolution of the average of the 12 wells in increase is of 3.16 mg/l per year, with like specific evolution between the two campaigns the contents of nitrates between 1992 and 2004 (Fig. 6):

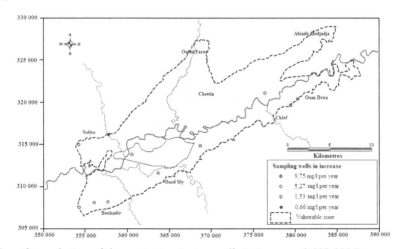

Fig. 6. Specific evolution of the contents nitrates wells in increase (1992-2004)

9.75 mg/l per year for 02 well, 5.27 mg/l per year for 03 wells, 1.53 mg/l per year for 06 wells, 0.66 mg/l per year for 01 well.

The excessive increases characterize Oum Drou (zone of horticulture) and the Boukadir downstream which coincides with the hydraulic downstream.

The global tendency of the evolution of the concentrations in nitrate of the waters of the alluvial aquifer of the middle western Cheliff translated a progressive temporal deterioration of the quality of these waters destined for the drinking and for the irrigation, since the percentage of well sampled during these last years offering non drinkable waters (> 50 mg/l) rose from 40% in the average to 63%.

3.1.5 Map of nitrates

Adjustment by a right of the points cloud between the measured values of nitrate and the residues (Fig. 7) show that the general tendency of the estimation is marked by a strong misjudgement of the values raised from NO_3; thus, all values that are superior or equal to 100 mg/l are underestimated systematically.

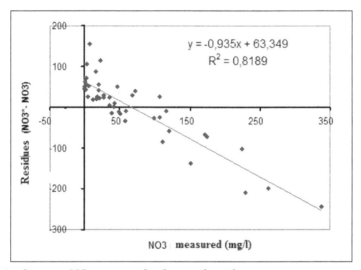

Fig. 7. Relation between NO_3 measured values and residues

3.1.5.1 Ordinary kriging

The experimental middle variogram was calculated on a distance of 28000 m without reaching a range (Fig. 8a). This variogram was adjusted with a linear model of 5300 of nugget with the same order of magnitude as the variance, thus translating a very high local variability (Douaoui et al., 2006).

It appears according to the map of nitrate established by ordinary kriging (OK method) (Fig. 8b) that the most affected zones are those for which the level of intensification of the N-fertilization (zones of garden farming) are the strongest (township of Sobha, Boukadir downstream, the southeasterly extension (to the west of Oum Drou), the plain of Medjadja) under the old alluviums and soils of borders area study characterized by the strongest permeabilities (10 cm/h). This strong hydraulic conductivity has for consequence that the transportation of waters of infiltration toward the deep layers makes itself very quickly

(Rahman, 2008; Giroux, 2003) and the sensitivity of these soils to the phenomenon of washing of the nitrates is therefore very elevated. In these same zones, the practices of breeding are more intensive

The map shows also a general bottom lower to 50 mg/l characterizing the centre of the area study. Soils in this zone have fine texture and weak permeability varying between 0.5 and 2 cm/h. the thick clayey profile that surmounts the aquifer in this zone and the weak yearly infiltration (25 mm) recorded in general in the semi-arid zones, seem to play an important role in this sense (Bettahar et al., 2009). In this same part, arboriculture concentrated in this part of area study is irrigated from dams waters of which the concentrations in nitrate are weak.

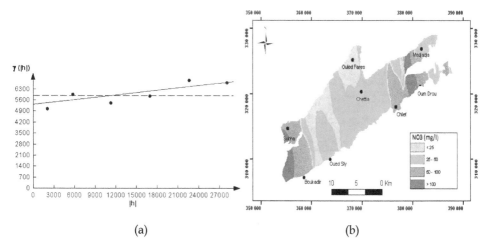

(a)   (b)

Fig. 8. Variogram and Map of nitrate concentrations estimated by Ordinary kriging (OK) method

3.1.5.2 Indicator kriging

The map established by Indicator Kriging (IK) method (Fig. 9) shows that the geographical distribution of the classes 50-100 and > 100 mg/l is generally the same that the one gotten by OK method. However, we observe an improvement in the elevated value surfaces (the class 50-100 mg/l) to the profit of those of the values excessively elevated (the class > 100 mg/l), weakly of the middle values (the class 25-50 mg/l) and even of the weak values (the class < 25 mg/l) in the low valley of the Ouahrane wadi. This zone is known by a strong agricultural activity (zone of garden farming, potato in particular benefitting from a phenomenal N-fertilization).

3.1.5.3 Comparison between the OK and IK methods

The quality of the estimation by the two types of kriging rests on the comparison between the surfaces estimated by every type (Fig. 10). The surfaces of the nitrate classes gotten by IK method compared to the OK show a reduction in the non contaminated surfaces and in the same way an increase of the surfaces very contaminated

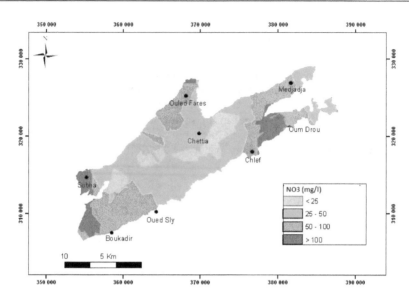

Fig. 9. Map of nitrate concentrations estimated by Indicator kriging (IK) method

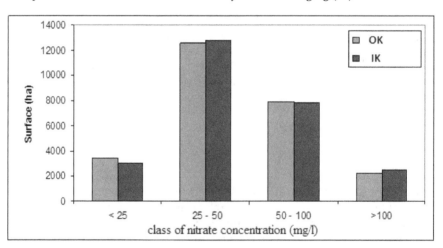

Fig. 10. Comparison between the surfaces of NO_3 classes gotten by OK and IK methods

3.2 Quantification of the nitrogen contributions

3.2.1 Contributions from N-fertilizers

The industrial chemical fertilizers, particularly, the NPK 15.15.15 is predominant for the quasi - totality of the exploitations with yearly middle doses of 500 kg ha⁻¹ for arboriculture and until 1000 kg ha⁻¹ for the potato.

The uses of other N-fertilizers as the urea (46%) and the sulphate of ammonium (21%) are estimated as high as 50-600 kg ha⁻¹ for the cereals, arboriculture and the garden farming.

The quantity of nitrogen gotten for every type of culture (Fig. 11) is deducted of the product of the dose of fertilizer that it receives by the corresponding surface.

3.2.2 Contributions from the water of irrigation

The surfaces of the garden farming and cereals are irrigated from the waters of wells of which NO3--N concentrations exceed the potability standard of 50 mg/l (Martin, 2003).

The total quantity of nitrogen brought by the water of irrigation represents only 3% of the one produced by the N-fertilizers (Fig. 11).

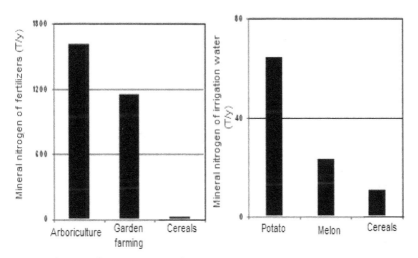

Fig. 11. Annual mineral nitrogen contribution

3.2.3 Contributions from breeding

The exploitations of the breeding for the different animal species (bovines, ovine, goats and poultries) are located in the borders of the valley (in the townships of Ouled fares, Abiadh Medjadja, Sobha and Boukadir). The calculation of the yearly total quantities of organic nitrogen generated by the set of every animal category is based on the values of nitrogen produces annually by head for every species, proposed by the CORPEN (Parris, 1998].

3.2.4 Contributions from municipal wastewater

Organic nitrogen estimated for Individual septic tank systems constitutes only 5% of the one generated by the breeding (Fig. 12).

3.2.5 Total contributions in nitrogen

Nitrogen brought by agriculture (fertilizers and water of irrigation) constitutes 86% of the total nitrogen brought to the soils of the valley. 97% of this last is attributed to nitrogenous fertilizers used extensively in garden farming, potatoes in particular. Extrapolated to the total irrigated area, this contribution is estimated at 238 kg ha^{-1} yr $^{-1}$.

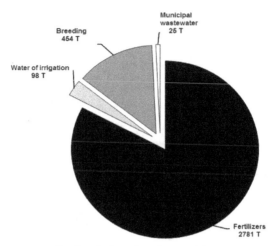

Fig. 12. Annual nitrogen contribution (mineral and organic)

3.3 Effect of the climate and soil characteristics on the nitrogen balance

The nitrogenous balance method proposed by the COMIFER (1996) and the CORPEN (1988), permits the nitrogenous excess calculation whose general formula can be given by the following equation:

$\sum ENTRIES - \sum EXITS$ = natural Contributions + non natural contributions - A - V - D - L
With A=Absorption by the plants, V= volatilization, D=dénitrification and L = Leaching.

3.3.1 The entries

3.3.1.1 Natural contributions

a.   Atmospheric nitrogen contributions: Nitrate concentration of the precipitations falling on the study area doesn't pass 2 mg/l (Ikhlef, 2008). This weak concentration doesn't seem to influence the entries.
b.   Contributions by mineralization: The organic matter rate is very weak on the soils of the study area (< 2%); this last will continue to decrease in the time in parallel with an increase of the speed of mineralization by effect of the semi-arid climate. The two phenomena decrease the capacity of soil to provide nitrogen by mineralization.

3.3.1.2 Non natural contributions

They constitute the sum of contributions from N-fertilizers, the water of irrigation, the breeding and municipal wastewater. They are valued to 3358 T yr $^{-1}$.

3.3.2 The exits

3.3.2.1 Absorption by the plants

If we keep the lower doorstep of the absorption rate (60%) of nitrogen contained in fertilizers recognized by Tremblay et al., (2001), the quantities of nitrogen absorbed by the plants in study area from the different fertilizers applied would be the order of 1669 T yr $^{-1}$.

3.3.2.2 Volatilization

The losses by volatilization depend on conditions of soil (pH, capacity of exchange, porosity, humidity) and of the climatic conditions (Sommer et al., 1991).

The losses by volatilization at the time of the application can reach 40 to 50% of nitrogen applied in the conditions of chalky soil, of pH> 7.5 and of elevated temperature (Tremblay et al., 2001; Hargrove, 1988). However, the urea remains the fertilizer that frees the strongest quantities of ammonia in the atmosphere, producing 72% of the quantities freed by the fertilizers (Environnement Canada, 2000). In the study area, the quantity of nitrogen that could be volatilized from the urea, for a doorstep of 40%, would be the order of 307 T yr $^{-1}$, either about 11% of the total of nitrogen (2781 T).

3.3.2.3 Denitrification

The fraction of nitrogen lost by denitrifiocation given N_2O is located particularly on soils badly aired to basic pH, in the conditions of elevated temperature (> 15°C). For a middle doorstep of 20% advanced by the works of Trembley and al. (2001) the fraction of applied mineral nitrogen capable to be topic to the denitrification in the study area would be meadows of 556 T yr $^{-1}$.

3.3.3 Effect of the climate and soil characteristics in semi-arid regions

In the semi-arid regions with strong agricultural activity, it is necessary to take always into account some entries: nitrogen brought by fertilization, nitrogen brought by the water of irrigation whatever weakly (Bettahar et al., 2008). It is generally useless to include nitrogen coming from the precipitations or the mineralization.

In the same way, it is indispensable to take into account some exits in the calculation of the nitrogenous balance: the quantity of nitrogen absorbed by the culture, the denitrification and the volatilization. The stape of calculation of the nitrate quantity leached is not always essential since the infiltration is very weak in these regions.

The risks of nitrate pollution in aquifers of the valleys in semi-arid climate seem bound closely to the climatic conditions and soil characteristics. Indeed, important quantities of nitrogen brought annually to the soils of Western middle Cheliff valley by different practices (agriculture, breeding and municipal waste water), don't reach the aquifer because of the climate and the soil characteristics.

The contribution of nitrogen by mineralization is weak, because of the reduction in the time, of the organic matter of soil in parallel with an increase of the speed of mineralization by effect of the semi-arid climate. The quantities of nitrate leaching in the aquifer, deducted of the nitrogenous excess, remain weak because of the weak yearly refill of the aquifer (25 mm only), direct consequence of the semi-arid climate of the study area and to the nature of soils whose hydraulic conductivity is weak (0,2 - 0,5 cm/h) on big surfaces of the valley.

4. Conclusion

Some knowledges have been acquired on the answer of the soils of Western middle Cheliff valley to the contributions of nitrogen coming from different origins. Indeed, the soils of borders, show a vulnerability more raised to the leaching of the nitrates seen their elevated permeability and that oscillates around 10 cm/h. Indeed, the nitrate concentrations in the aquifer are there the most elevated (> 100 mg/l).

On the contrary, the soils of the plain (center of the valley) seem less vulnerable to the nitrate leaching; they have fine texture and weak permeability varying between 0.5 and 2 cm/h. It could attenuate the propagation of the nitrates strongly in depth. The thick clayey profile that surmounts the aquifer, in this part of the valley seems to play an important role also in this sense.

The chemical characteristics of the soils of the valley, notably the pH and the rate of limestone, can strongly encourage important losses of nitrogen through processes as the volatilization and the denitrification. These last can appear very increased seen the temperatures excessively elevated in the Cheliff plains in the months of August and September, period of irrigation of the garden farming very developed on the borders of the valley, potato in particular, by the well waters greatly loaded in nitrate.

In the semi-arid regions characterised by a strong agricultural activity, it is necessary to take always into account some entries: nitrogen brought by fertilization, nitrogen brought by the water of irrigation whatever weakly. It is generally useless to include nitrogen coming from the precipitations or the mineralization.

In the same way, it is necessary to take into account some exits in the calculation of the nitrogenous balance: the quantity of nitrogen absorbed by the culture, the denitrification and the volatilization particularly for the chalky soils with a basic pH. The calculation of the quantity of nitrates leaching is not always essential since the infiltration is very weak in these regions.

Although the risks of nitrate pollution in the aquifers of the valleys in semi-arid climate seem lessen, even with phenomenal contributions of nitrogen (3000 T annually in the case of the study area), because of the climate and of the physical and chemical characteristics of soil, these waters are not completely safe from nitrate contamination.

5. References

Addy K.L., Gold A.J., Groffman P.M. et Jacinth P.A. (1999). Ground water nitrate removal in subsoil of forested and mowed riparian buffer zones. *J. Environ. Qual.*, 28, 962-970, ISSN 1537-2537

Benoit M., Bonneau M. et Dambrine E. (1997). Influence du sol et de sa mise en valeur sur la qualité des eaux infiltrées et superficielles. *L'Eurobiologiste*, Tome XXXI, (230), 53-58, ISSN 0999-5749

Bettahar N., Kettab A., Ali Benamara A. et Douaoui A. (2008). Effet des conditions pédoclimatiques sur le bilan d'azote. Cas de la vallée du moyen Cheliff occidental. *Algerian Journal Of Technology -AJOT, ISSN 1111-357X, Number Special – An International Publication of Engineering Sciences*, 1, 441-447.

Bettahar N., Ali Benamara A., Kettab A. et Douaoui A. (2009). Risque de pollution nitratée dans les zones semi-arides. Cas de la vallée du moyen Cheliff occidental. *Rev. Sci. Eau*, 22(1), 69-78, ISSN 0298-6663

Comifer (1996). *Calcul de la Fertilisation Azotée Des Cultures Annuelles*. Comité Français d'étude et de développement de la Fertilisation Raisonnée.

Corpen (1988). *La fertilisation raisonnée.* Comité d'Orientation pour des Pratiques agricoles respectueuses de l'Environnement.

Delgado J.A. et Shaffer M.J. (2002). Essentials of a national nitrate leaching index assessment tool. *Journal of Soil and Water Conservation*, 57, 327-335, ISSN 1941- 3300

Douaoui A., Herve N. et Walter CH. (2006). Detecting salinity hazards within a semiarid context by means of combining soil and remote-sensing data. *GEODERMA*, 134, 217-230, ISSN 0016-7061

Elmi, A.A., Madramootoo C., Egeh M. et Hamel C. (2004). Water and fertilizer nitrogen management to minimize nitrate pollution from a cropped soil in south western Quebec. Canada. *Water Air Soil Pollut.*, 151, 117-134, ISSN 1573-2940

Environnement Canada (2000). Canadian Environmental Protection Act-Priority Substances List-*1995 Ammonia Emissions Guidebook. Première ébauche.* Direction des données sur la pollution, Hull (Québec).

Feng, Z.Z., Wang, X.K. et Feng, Z.W. (2005). Soil N and salinity leaching after the autumn irrigation and its impact on groundwater in Hetao Irrigation District, China. *Agric. Water Manage*, 71, 131-143, ISSN 0378-3774

Firestone M.K. (1982). Biological denitrification. In *"Nitrogen in agricultural soils"* (F. J. Stevenson, ed). Am; Soc. Agron., Madison, Wisconsin, pp 289-318.

Giroux I. (2003). *Contamination de l'eau souterraine par les pesticides et les nitrates dans les régions en culture de pommes de terre.* Direction du suivi de l'état de l'environnement. Ministère de l'Environnement. Québec, 23 p.

Gomez E. (2002). *Modélisation intégrée du transfert de nitrate à l'échelle régionale dans un système hydrologique. Application au bassin de la seine.* Thèse Doct. Ecole des Mines de Paris, 218 p.

Hargrove W.L. (1988). Evaluation of ammonia volatilization in the field. *J. Prod. Agri.*, 1, 104-111, ISSN 0890-8524

Haynes R.J. (1986b). Uptake and assimilation of mineral nitrogen by plants. In *Mineral nitrogen in the plant-soil system,* pp. 303-378, Physiological ecology. Ed. TT Kozlowsky, Madison, Wisconsin, 483 p.

Hénault C. et Germon J.C. (1995). Quantification de la dénitrification et des émissions de protoxyde d'azote (N_2O) par les sols. *Agronomie*, 15, 321-355, ISSN 1773-0155

Ikhlef S. (2008). *Etude de la pollution de la nappe alluviale du haut Cheliff par les nitrates.* Mémoire de Mag., Univ. HBB Chlef, 145 p.

Machet J.M., Pierre D., Recours S. et Remy J.C. (1987). Signification du coefficient réel d'utilisation et conséquence pour la fertilisation azotée des cultures. *C. R. Acad. Agric. De France*, 3, 39-55, ISBN/ISSN 0989-6988

Martin C. (2003). *Mécanismes hydrologiques et hydrochimiques impliqués dans les variations saisonnières des teneurs en nitrate dans les bassins versants agricoles. Approche expérimentale et modélisation.* Thèse Doctorat, Ecole de Renne 1- France, 269 p.

Mattauer M. (1958) *Etude géologique de l'Ouarsenis oriental (Algérie).* Thèse Es sciences, Besançon, France, 343 p.

Parris K. (1998). Agricultural nutrient balances as agri-environmental indicators: an OECD perspective. *Environmental Pollution*, 102, 219-225, ISSN 0269-7491

Payraudeau S. (2002). *Modélisation distribuée de flux d'azote sur des petits bassins versants méditerranéens.* Thèse Doctorat. ENGREF- Montpellier, 245 p.

Perrodon A. (1957) *Etude géologique des bassins néogènes sublittoraux de l'Algérie Nord Occidentale*, Thèse de Doctorat, 115 p.

Pinheiro A. (1995). *Un outil d'aide à la gestion de la pollution agricole: le modèle POLA*. Thèse Doctorat, INP – Toulouse, 344 p.

Rahman A. (2008). A GIS based DRASTIC model for assessing groundwater vulnerability in shallow aquifer in Aligarh, India. *Applied Geography*, 28, 32-53, ISSN 0143-6228

Rassam D.W., Pagendam D.E. et Hunter H.M. (2008). Conceptualisation and application of models for groundwater-surface water interactions and nitrate attenuation potential in riparian zones. *Environmental Modelling & Software*, 23, 859-875, ISSN 1364-8152

Scet Agri (1984). *Bilan des ressources en sol. Etude du réaménagement et de l'extension du périmètre du moyen Chéliff*: Rap A1.2. 1. Pub. Ministère de l'Hydraulique, Algérie, 35p.

Sivertun A. et Prange L. (2003). Non-point source critical area analysis in the Gisselo-watershed using GIS. *Environmental Modelling&Software*, 18(10), 887-898, ISSN 1364-8152

Smith M.S. et Tiedge J.M. (1979). Phases of denitrification following oxygen depletion in soil. *Soil Biol. Biochem.*, 11, 261-267, ISSN 0038-0717

Sommer S.G., Olesen J.E. et Christensen B.T. (1991). Effects of temperature, wind speed and air humidity on ammonia volatilization. *Journal of Agricultural Sciences*, 117, 91-100, ISSN 1916-9760

Standford G., Vander Pol R.A. et Dzienia S. (1975). Denitrification rates in relation to total and extractable soil carbon. *Soil. Sci. Soc. Amer. Proc.*, 39, 284-289, ISSN 0038-0776

Tremblay N. Scharpf H.C., Weier U., Laurence H. et Owen J. (2001). *Régie de l'azote chez les cultures maraîchères, Guide pour une fertilisation raisonnée*, Agriculture et Agroalimentaire, Canada, 70 p.

Van Bol V. (2000). *Azote et agriculture durable. Approche systématique en fermes-pilotes*. Thèse Doctorat de l'Université catholique de Louvain, 157 p.

Weier U. (1992). Effect of splitting N fertilizer and yield of broccoli. Versuche in *Deutschen Gartenbau*. N° 26.

Part 2

Plant Breeding

3

Assessment of Diversity in Grapevine Gene Pools from Romania and Republic of Moldova, Based on SSR Markers Analysis

Ligia Gabriela Ghețea et al.*
University of Bucharest, Faculty of Biology, Bucharest,
Romania

1. Introduction

In the last 15 years, inventory of the genetic diversity for agricultural purposes and environmental protection has become a constant preocupation of European Union in the biological research field.

Due to the economical importance of *Vitis vinifera* species, and to the long viticulture tradition of many countries, the European Union has developed international projects having as main objective improving the knowledge, conservation and sustainable use of *Vitis* genetic resources in Europe.

One of the first such initiatives was GENRES CT96 081 (European Network for Grapevine Genetic Resources Conservation and Characterization) project (started in March 1997, ended in February 2002), aiming to establish the European *Vitis* Database, with free access via Internet, in order to enhance the utilization of relevant and highly valuable germplasm in breeding. The *Vitis* Working Group which has been constituted at the end of the GENRES project, decided to start the establishment of an SSR-marker database as part of the European *Vitis* database (www.ecpgr.cgiar.org/workgroups/vitis/).

Another international project, financially supported by the Government of Luxembourg and coordinated by Biodiversity International, named „*Conservation and sustainable use of grapevine germplasm in Caucasus and Northern Black See region"*, was developed during the years 2003-2008, having as main result a better conservation of germplasm collections in this region (Armenia, Azerbaijan, Georgia, Moldova, Russian Federation and Ukraine) which represent a very unique and rich source of grapevine genetic variation.

*Rozalia Magda Motoc[1], Carmen Florentina Popescu[2], Nicolae Barbacar[3], Ligia Elena Bărbării[4],
Carmen Monica Constantinescu[4], Daniela Iancu[4], Tatiana Bătrînu[3], Ina Bivol[3], Ioan Baca[3]
and Gheorghe Savin[5]
[1]University of Bucharest, Faculty of Biology, Bucharest, Romania
[2]National Institute of Research & Development for Biotechnologies in Horticulture, Ştefăneşti-Argeş, Romania
[3]Genetics and Plant Physiology Institute of the Moldavian Academy of Sciences, Chişinău, Republic of Moldova
[4]National Institute of Legal Medicine „Mina Minovici", Bucharest, Romania
[5]National Institute for Viticulture and Oenology, Department of Grapevine Genetic Resources, Chişinău,
Republic of Moldova*

Starting from 2010, a third important European project is in course (ending in 2014): the COST project named „*East-West collaboration for grapevine diversity exploration and mobilization of adaptive traits for breeding*". In this project are involved 35 member states countries (Austria, Belgium, Bosnia and Herzegovina, Bulgaria, Croatia, Cyprus, Czech Republic, Denmark, Estonia, Finland, France, Germany, Greece, Hungary, Iceland, Ireland, Italy, Latvia, Lithuania, Luxembourg, Malta, The Netherlands, Norway, Poland, Portugal, Romania, Slovakia, Slovenia, Spain, Sweden, Switzerland, Turkey, United Kingdom, Serbia, Former Yugoslav Republic of Macedonia), one cooperating state Israel, and also reciprocal agreements with Australia, New Zealand, South Africa and Argentina. The main objective is to create a knowledge platform about European grapevine germplasm, essential for (1) the maintenance of genetic resources, (2) the discrimination and identification of grapevine varieties and (3) the availability and exchange of germplasm. These grapevine collections gathering autochthonous and unique varieties with valuable genetic- and phenotypical traits, are also essential for long-term conservation and sustainable use of this economically important species.

Romania and Republic of Moldova have a multimillenary tradition in grapevine cultivation and wine production. Thus, both ancient- and newly created grapevine varieties from these two countries are valuable gene resources which must be inventored for a complete genetic characterization, based on reliable molecular markers.

As a response to the European Union initiative for inventory and conservation of grapevine genetic resources, two research teams from Romania and Republic of Moldova have initiated a research direction aiming to establish a genetic profile – based on SSR markers analysis - for the grapevine varieties cultivated in these two countries, in order to make an inventory of them, and also to facilitate the registration of Romanian- and Moldavian originated cultivars in the European *Vitis* Database. These specific genetic profiles represent also a real passport that certifies their authenticity and represents a guarantee for further preservation of grapevine cultivars with economic value.

2. SSR markers for designing the genetic profile of grapevine cultivars

SSR (*s*imple *s*equence *r*epeats, or microsatellites) are arrays of short motifs, highly repeated, of 1 to 6 base pairs. These locus specific markers are characterized by hypervariability, abundance, high reproducibility, Mendelian inheritance and codominant nature. They are not affected by environmental factors such as soil, climate, cultivation methods or diseases (Scott et al., 2000; Meredith, 2001).

Therefore, microsatellites are the favourite type of DNA markers used to characterized the grapevine cultivars, their properties enabling a wide range of applications: cultivar identification and discrimination, parentage testing and pedigree reconstruction, genetic stability of all morphological-physiological features, yield capacities of each cultivar, management of germplasm collection (Thomas & Scott, 1993; Bowers et al., 1996; Thomas et al. 1998; Bowers et al., 1999; Sefc et al., 1999; Sefc et al., 2000; Sefc et al., 2001). Their relative abundance within the genome and being very easy to be detected by PCR reaction explain that at present, more than 600 grapevine microsatellite markers are available (Moncada et al., 2006), making possible refined genetic profiles of different varieties.

Analysis of microsatellite loci in different cultivars (genotypes) provides useful information about: (i) genetic data for each conserved genotype; (ii) taxonomy relatedness, geographic origin and ecological aspects (genetic diversity); (iii) the distinctness between cultivars, uniformity of planting material belonging to the same genotype, and genetic stability of all morphological-physiological features and yield capacities of each cultivar.

The *Vitis* Working Group highly recommended to include in each SSR-marker genotyping in grapevine cultivars, six microsatellite loci which proved to have a high variability among different cultivars, making possible a good discrimination even between very close related ones. These SSR markers are: VrZAG62, VrZAG79, VVS2, and VVMD5, VVMD7, VVMD27. In our researches, we obtained good results with five of these SSR markers, along with other ten microsatellite loci.

From our work during the 7 past years, on grapevine cultivars genotyping based on SSR markers, for this book chapter we have made a selection of the most interesting results, to be presented. Some of these results have already been published (Gheţea et al., 2010, a; b), some others are in course to be published.

3. The grapevine cultivars for which genetic profiles have been obtained

The grapevine cultivars considered for the study are among the most economically important, and encountered in the vineyards, in the two countries. The Romanian grapevine cultivars are included in the Official Catalogue of Cultivated Plants from Romania, and are maintained in the stock collection of the National Institute for Research and Development for Biotechnologies in Horticulture (NIRDBH), Ştefăneşti-Argeş. The Moldavian grapevine cultivars are in the stock collection of the National Institute for Viticulture and Oenology, Chişinău.

In this paper, we present our results on 14 grapevine cultivars, 7 cultivars from each of the two stock collections.

3.1 The grapevine cultivars from Romanian stock collection

Although, in Romania, vineyards produce grapes of unsurpassed quantity and quality with autochthonous genotypes, these cultivars are known and appreciated only in few regions. Also, the new breeders' creations are planted only in local areas and are commercialized mostly on national market. As Romanian grapevine cultivars were not studied from the genetic point of view, very limited information is available about their origin and gene pool value. Such genetic data regarding both the ancient and new creations of grapevine varieties from Romania become essential for recognition and registration in the International Grapevine Genome database. Moreover, this must be the main reason to promote a sustained research activity for a complete evaluation of Romanian grapevine genetic resources. So, the first step was to establish a core colection and a research program with the following objectives:

a. to verify the genetic identity of the cultivars, their stability and integrity during producing and conservation;
b. to identify duplicates in the grapevine collection, to eliminate them from the collection and to guaranty the authenticity of the planting material;
c. to prove the distinctness of the ancient and new grapevine varieties, important requirement for characterization and registration of Romanian germplasm.

d. Expectations as results from these activities are: a) a complete view of the Romanian grapevine gene resources; b) improve knowledge of the patent of distribution of grapevine genetic diversity; c) the genetic characterization of the Romanian cultivars would be equally very useful for all research units, for the owners of genotypes, for the grapevine growers and wine makers.

The cultivars we are presenting are as follows: (Dobrei et al., 2005):

1. **Italian Riesling** – opinions concerning its origin are different: German from Rhine Valley, or Austrian from Styria region, or Italian origin.
2. **Sauvignon** – French origin; with large areas cultivation; extensively cultivated in Bordeaux region.
3. **Cabernet Sauvignon** – French origin, old cultivar obtained in Bordeaux region, by cross-breeding between *Cabernet Franc* and *Sauvignon Blanc*; high geographical dispersion.
4. **Tămâioasă Românească** – uncertain origin; very old cultivar, known in ancient Greece as "Anathelion moschaton", and also during in Roman Empire, as "Aspinae"; cultivated in Romanian vineyards from ancient times, it is considered as autochthonous cultivar.
5. **Fetească neagră** – ancient Romanian cultivar, obtained by empirical selection from *Vitis sylvestris* Gmel; originated on the Prut river, in Iasi region.
6. **Fetească albă** – considered as autochthonous cultivar, obtained by empirical selection, from *Fetească neagră* cultivar; extensively cultivated in Romanian vineyards.
7. **Fetească regală** – Romanian origin cultivar, it is supposed to be result of a natural hybridization obtained by cross-breeding between two autochthonous cultivars: *Fetească albă* and *Grasă de Cotnari*.

3.2 The grapevine cultivars from Moldavian stock collection

A constant preoccupation in Moldavian breeding programs in viticulture, during the last 40 years, was obtaining valuable grapevine varieties, including large bunches and berries, early maturation of fruits, high accumulation of sugar, seedless berries, resistance of unfavourable environmental factors (drought, harsh winter conditions, diseases and pests) (Savin et al., 2010).

The cultivars we are discussing here are as follows:

1. **Cabernet Sauvignon** – its origin is mentioned above (at 3.1. section).
2. **Fetească neagră** – Romanian origin (see above).
3. **Fetească albă** – Romanian origin; extensively cultivated in Moldavian vineyards (see previous section).
4. **Fetească regală** – Romanian origin (see previous section).
5. **Muscat timpuriu de București** – table grapes with early ripening, Romanian origin; obtained by cross-breeding between Romanian cultivars *Coarnă albă* and *Regina Viilor*.
6. **Timpuriu de Cluj** – table grapes, Romanian origin; obtained by cross-breeding between two Romanian cultivars: *Crâmpoșie* and *Frumoasă de Ghioroc*.
7. **Apiren extratimpuriu** – seedless cultivar obtained at National Institute for Viticulture and Oenology, Chișinău, by cross-breeding between *Urojainyi* and *Kismis turkmenski* cultivars; valuable for early ripening, high fertility and resistance to harsh winter conditions.

4. Methodology

DNA was extracted from young leaves of different grapevine genotypes, according to the method reported by Vallejos et al. (1992). For genotyping the selected cultivars, primer pairs for 15 microsatellite loci were chosen, 5 of which being recommended as reference SSR markers by the *Vitis* Working Group (www.ecpgr.cgiar.org/workgroups/vitis/).

The forward primer of each pair has been marked with one of the four fluorochromes: 6-FAM, NED, VIC or PET. The liophylised primers have been obtained from Applied Biosystems, USA.

For the amplification reaction, a thermocycler (Bio-Rad) with Peltier system was used. Optimal primer annealing temperature, presented in Table 1, was chosen for each primer pair, according to the results obtained in the temperature gradient PCR.

The PCR mix have had the following composition: 5X Colorless GoTaq® Flexi Buffer (Promega) – 5μl; MgCl$_2$ – 0,75μl; dNTP mix, 10mM (Promega) – 0,5μl; primer 1 – 1μl; primer 2 – 1μl; DNA sample – 30-45 ng/μl; GoTaq® DNA Polymerase (Promega) – 0,25μl; ddH$_2$O – X μl (X was calculated for each sample, depending on the volume of DNA sample, to a final volume of 25μl).

The PCR reactions were performed on an GeneAmp PCR System 9700 thermocycler (Applied Biosystems), using for the annealing step, the optimal temperature established for each primer pair (51^{0}C, 52^{0}C, 54^{0}C or 56^{0}C) (see Table 1).

Locus	Allele size range (bp) cited in the literature	Primer	
		Sequence	T$^\circ_{annealing}$
ssrVrZAG7[1]	106 – 158	(F) gtggtagtgggtgtgaacggagtgg (R) aacagcatgacatccacctcaacgg	56°C
ssrVrZAG12[1]	140 – 172	(F) ctgcaaataaatattaaaaaattcg (R) aaatcctcggtctctagccaaaagg	51°C
ssrVrZAG15[1]	163 – 193	(F) ggattttggctgtagttttgtgaag (R) atctcaagctgggctgtattacaat	54°C
ssrVrZAG62[1]	185 – 203	(F) ggtgaaatgggcaccgaacacacgc (R) ccatgtctctcctcagcttctcagc	54°C
ssrVrZAG79[1]	236 – 260	(F) agattgtggaggagggaacaaaccg (R) tgcccccattttcaaactcccttcc	54°C
VVS1[2]	160 – 205	(F) acaattggaaaccgcgcgtggag (R) cttctcaatgatatctaaaaccatg	54°C
VVS2[2]	129 – 155	(F) cagcccgtaaatgtatccatc (R) aaattcaaaattctaattcaactgg	54°C
VVS4[2]	167 – 187	(F) ccatcagtgataaaacctaatgcc (R) cccaccttgcccttagatgtta	56°C
VVMD6[3]	194 – 214	(F) atctctaaccctaaaaccat (R) ctgtctaagacgaagaaga	52°C
VVMD7[3]	233 – 263	(F) agagttgcggagaacaggat (R) cgaaccttcacacgcttgat	54°C
VVMD14[3]	222 – 250	(F) catgaaaaaatcaacataaaagggc (R) ttgttacccaaacacttcactaatgc	54°C

VVMD17[3]	212 – 236	(F) tgactcgccaaaatctgacg	52°C
		(R) cacacatatcatcaccacacgg	
VVMD21[3]	243 – 266	(F) ggttgtcttatggagttgatgttgc	52°C
		(R) gcttcagtaaaaagggattgcg	
VVMD27[3]	173 – 194	(F) gtaccagatctgaatacatccgtaagt	54°C
		(R) acgggtatagagcaaacggtgt	
VVMD36[3]	244 – 315	(F) taaaataataataggggggacacggg	56°C
		(R) gcaactgtaaaggtaagacacagtcc	

([1] Sefc et al., 1999; Nuclear SSR primers of the Centre for Applied Genetics, University of Agriculture, Vienna, Muthgasse 18, A-1190 Vienna, Austria)

([2] Thomas & Scott, 1993; Thomas et al., 1998; Nuclear SSR primers of the Division of Horticulture CSIRO, Adelaide, Australia)

([3] Bowers et al.,1996; Bowers et al.,1999; Nuclear SSR primers of the Department of Viticulture and Enology, University of California, Davis, USA)

Table 1. List of the analysed SSR loci

The PCR reaction steps were:

I - $95^0C \rightarrow$ 4 minutes (initial denaturation step)
        $95^0C \rightarrow$ 1 minute | x 35 cycles
II - X^0C (X=51^0C, 52^0C, 54^0C or 56^0C) $\rightarrow$ 1 minute
        $72^0C \rightarrow$ 1 minute
III - $72^0C \rightarrow$ 7 minutes (final elongation step)

The efficiency of the amplification reaction was analysed in a 2% agarose gel, in TAE buffer, according to Ausubel et al. (1990). A 5µg/ml ethidium bromide solution was used for the analysis of amplicon bands in UV light.

The amplicon genotyping was performed at an ABI PRISM 3100 Genetic Analyzer, using ROX 500 as internal standard. The samples were analysed with the GeneMapper program.

5. Results and discussions

Genotyping method allowed us to determine the base pair number in each amplicon obtained for the 15 microsatellite loci. Different cultivars showed, at a certain SSR locus, a homozygotic (the presence of a single allelic variant), or a heterozygotic (two, and in some cases, even 3 allelic variants) condition (see Table 2 and Table 3).

In figure 1, an electropherogram obtained using the GeneMapper, on ABI PRISM 3100 Genetic Analyser, is presented (Gheţea et al., 2010, a).

Based on amplicon dimensions corresponding to the 15 SSR markers, a genetic profile was obtained for each of the analysed cultivars.

The **VrZAG7** locus showed a homozygous condition in all cultivars we have analysed. The allelic variants having the highest frequency in Romanian cultivars are those of 155bp and 157 base pairs (bp) (approximately 43%), while in the Moldavian ones, the most frequent (approximately 29%) is of 156bp (Table 4 and Table 5). Most of the detected alleles are within the size interval cited in the literature (see Table 1 for allelic size range); in 3 Moldavian provenience cultivars (*Fetească regală, Muscat Timpuriu de Bucureşti* and *Timpuriu* *de Cluj*), 2 significantly shorter allelic variants (of 102/103, and 107 bp) have been observed. The 102/103 variant is outside the known allelic size range (Table 2 and Table 3). The high degree of homozygocity showed at this SSR locus is an indication that it could be associated with a coding region where genes for economically important traits are located (Oliveira et al., 2006).

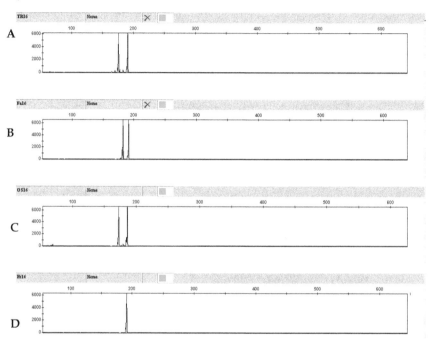

Fig. 1. Gene Mapper image for *Tămâioasă Românească* (A), *Fetească albă* (B), *Cabernet Sauvignon* (C) and *Fetească regală* (D) cultivars, at the VVMD27 microsatellite region. A heterozygotic condition appears for the first three cultivars, with significant diference (in base pair number) between the two allelic variants, while the fourth one is homozygous at this locus.

For the **VrZAG12** region, most of the analysed cultivars showed a homozygotic status. Only *Cabernet Sauvignon* and *Apiren extratimpuriu* from Republic of Moldova stock collection, and *Italian Riesling* from Romanian stock collection, are heterozygous. Significant differences appear in the same cultivar comming from different stock collections: *Fetească albă* Romania (R) – 150:150/Republic of Moldova (M) – 165:165; *Fetească regală* (R) – 170:170/(M) – 153:153; *Fetească neagră* (R) – 148:148/(M) – 166:166 (Table 2 and Table 3). One can conclude that this microsatellite locus could be placed in a coding region linked with genes for valuable traits. Although, a molecular mechanism (probably, during DNA replication) acting differently in the two gene pools (Romanian and/or Moldavian), by adding or remove a number of base pairs, created, in time, different allelic variants stabilized, at first, by self-pollination and later, by repeated artificial selection and vegetative multiplication. The most frequent allele at this SSR locus has 147/148bp, and is encountered in a homozygotic- or heterozygotic status, also in Romanian and Moldavian collection (Table 4 and Table 5). All the allelic variants we have found are within the allele size interval cited in the literature (Table 1).

The most frequent and well conserved allelic variant for the **VrZAG15** region, has 164 (163-165) bp (Table 4 and Table 5), and appears in a homozygotic status in 9 cultivars, and in a heterozygotic status, in other 4 cultivars (Table 2 and Table 3).

The **VrZAG62** locus was recommended by the *Vitis* Working Group as SSR marker due to high allelic variability observed in grapevine cultivars of different gene pools, in different countries. This high degree of variability is confirmed also by our results: in these 14 cultivars we discuss here, 5 allelic variants have been found in Romanian cultivars (the most frequent variant, of 198bp, approximately 36%), and 7 different alleles in Moldavian cultivars (the most frequent variant, of 190bp, approximately 50%) (Table 4 and Table 5). They are also in a homozygotic- or heterozygotic status (Table 2 and Table 3). In other Moldavian cultivars, we have obtained, at this locus, even 3 or 4 different amplicons, corresponding to a triallelic- or a tetraallelic condition (article in course to be published). This is not a singular SSR locus where multiallelic condition has been encountered, and this particular feature will be discussed later. The cultivars *Fetească albă*, *Fetească regală* and *Cabernet Sauvignon* have different allelic constitutions, depending on the Romanian- or Moldavian stock collection origin. Two allelic variants, of 205 and 206bp (slightly longer than the size interval cited in the literature), have been found in Moldavian cultivars *Fetească albă* and *Cabernet Sauvignon*. The SSR locus VrZAG62 was identified on the linkage group corresponding to chromosome 7 of the *Vitis vinifera* complement (Riaz et al., 2004; Lowe & Walker, 2006; Troggio et al., 2007; Vezzulli et al., 2008), probably in an non-coding region, which explains its high mutational rate (Oliveira et al., 2006).

Cultivar	Microsatellite loci														
	ZAG7	ZAG12	ZAG15	ZAG62	ZAG79	VVS1	VVS2	VVS4	VVMD 6	VVMD 7	VVMD 14	VVMD 17	VVMD 21	VVMD 27	VVMD 36
Fetească albă R	157	150	172 174	194	-	178 183	**128**	168	200 207	248 254	**216** 228	**210** 219	244	182 191	261 271
Fetească albă M	156	165	163	200 **206**	246 250	178	131 148	177	198 207	240 250	232	221	245 255	179	**240** 266
Fetească regală R	158	170	164 172	196	246	177	**128**	168	**187** 197	248 250	**216**	**209** 219	245	191	249 261
Fetească regală M	107	153	165 192	190 196	-	179 188	129	169 177	207	250 252	**219** 225	219	247 253	191	264 272
Fetească neagră R	157	148	162	198	246 254	176 186	139	167	**186**	240 256	**216** 228	**210** 219	**239** 244	176 186	249 259
Fetească neagră M	156	166	163	190 196	250 254	179 187	**126**	-	200 207	240	236	220	-	177 191	**236** **240**
Cabernet Sauvign. R	155	148	162 164	190 196	**228** 247	178	135 148	169 176	200	241	**218** 222	218	245 255	**172** 186	249 259
Cabernet Sauvign. M	157	147 156	164	190 **205**	248 258	170 178	**128** 141	176	**173** **187**	244	**218** 228	217 231	243	180 186	**103** **237** 262

Table 2. Allelic size (number of base pairs) at 15 microsatellite loci, in 4 grapevine varieties cultivated also in Romania and Republic of Moldova (see section 3.). *R* – Romania; *M* – Republic of Moldova. Bolded numbers – allelic variants shorter than the allele sizes cited in the literature (see Table 1). Dashes – no amplicon has been obtained.

Cultivar	Microsatellite loci														
	ZAG7	ZAG12	ZAG15	ZAG62	ZAG79	VVS1	VVS2	VVS4	VVMD 6	VVMD 7	VVMD 14	VVMD 17	VVMD 21	VVMD 27	VVMD 36
Italian Riesling R	157	147 140	164 174	198	**230**	180	130 148	168	200 207	248 258	**216**	219	243	182 186	249 259
Sauvignon R	155	148	164	190 196	**235**	178	**128** 148	169 176	**191** 200	241 258	**216**	220	246 253	182 191	267
Tămâioasă Română nească R	155	150	164	188 198	248 252	176	**128**	169 176	199	235	**216**	219	244 263	176 191	**239** 259
Muscat Timpuriu de Buc. M	**102**	146	164	190	238 253	178 185	141	169 176	206	240	**217**	219	245 252	178	259
Timpuriu de Cluj M	**103**	146	164	190	237	178	148 153	176	206	240	**217**	220	246	176 182	260
Apiren extratimpuriu M	158	147 155	164	189 195	237 248	178	139 148	176	206 212	240 252	-	218 220	245 253	182	**103** **234** **236**

Table 3. Allelic size (number of base pairs) at 15 microsatellite loci, in 6 grapevine varieties cultivated in Romania or Republic of Moldova (see section 3.) R – Romania; M – Republic of Moldova. Bolded numbers – allelic variants shorter than the allele sizes cited in the literature (see Table 1). Dashes – no amplicon has been obtained.

The **VrZAG 79** locus was identified on the linkage group corresponding to chromosome 5 of the *Vitis vinifera* complement (Riaz et al., 2004; Troggio et al., 2007; Vezzulli et al., 2008). Like VrZAG62 locus, this region has a high variability of the number of base pairs, generating many allelic variants: in these 14 cultivars, we have found 8 different alleles also in Romanian- and in Moldavian group; the most frequent allelic variant in Romanian cultivars has 246bp (approximately 25%), and in Moldavian ones – 235bp – (approximately 25%) (Table 4 and Table 5); 3 allele forms – those of 228bp, 230bp and 235bp – are outside the allelic size range known in the literature (they are slightly shorter). Interesting, the alleles of 228bp and 230bp have been found in 3 non-Romanian origin cultivars: *Cabernet Sauvignon* (228:247), *Italian Riesling* (230:230) and *Sauvignon* (235:235), existent in Romanian stock collection (Table 2 and Table 3). This aspect indicates that the 3 cultivars were imported in the Romanian grapevine gene pool a long time ago, suffering discrete genetic modifications, under the influence of local molecular mechanisms.

For the **VVS1** microsatellite locus, homozygotic- and also heterozygotic cultivars have been found; the most frequent allelic variant was that of 178bp, in both Romanian and Moldavian group. The frequency of this allele in the 14 analysed cultivars is of 36% in Romanian group, and of 57% in Moldavian group (Table 4 and Table 5). At this locus, like in other SSR loci, some differences appear in the same cultivar provenant from different stock collections (Romanian / Moldavian): *Fetească albă* (R) – 178:183/(M) – 178:178; *Fetească regală* (R) – 177:177/(M) – 178:188; *Cabernet Sauvignon* (R) – 179:179/(M) – 170:178 (Table 2 and Table 3). All the allelic variants found in the 14 cultivars, are within the allelic size interval (Table 1).

The **VVS2** locus was identified on the linkage group corresponding to chromosome 11 of the *Vitis vinifera* complement (Riaz et al., 2004; Lowe & Walker, 2006; Troggio et al., 2007; Vezzulli et al., 2008). This locus was selected by the *Vitis* Working Group as SSR marker, due to its high allelic variability. In the 14 cultivars, we found 5 different alleles in Romanian group (most frequent being that of 128bp – 50%), and 8 allelic variants in Moldavian group (most frequent being those of 141bp and 148bp – 21,5% each of them) (Table 4 and Table 5); 2 allelic variants – those of 128bp, and 126bp – are shorter than the allelic size interval known for this locus (Table 1). The allele of 148bp is encountered both in Romanian- and Moldavian stock collection. The allelic variant of 128bp was found in an homozygotic condition in 3 Romanian cultivars: *Fetească albă*, *Fetească regală*, and *Tămâioasă Românească*. *Cabernet Sauvignon* from Moldavian collection has also the 128bp allele, but is heterozygous at this locus (128:141). 3 allelic variants, of 141bp, 129bp and 126bp were encountered only in Moldavian provenance cultivars *Muscat Timpuriu de București* (141:141), *Cabernet Sauvignon* (128:141), *Fetească regală* (129:129), and *Fetească neagră* (126:126) (Table 2 and Table 3).

The **VVS4** locus was identified on the linkage group corresponding to chromosome 8 of the *Vitis vinifera* complement (Riaz et al., 2004; Troggio et al., 2007). This SSR region revealed a moderate allelic variability. In Romanian cultivars 4 allelic variants have been detected (with the most frequent, that of 168bp – 43%), while in the Moldavian group only 3 different alleles have been found (with the most frequent, having 176bp – 58%) (Table 4 and Table 5). The 168/169 allelic variants are well conserved in Romanian collection provenance cultivars, also in an homozygotic status (*Fetească albă*, *Fetească regală*, *Italian Riesling*), and also in an heterozygotic one (*Sauvignon* and *Cabernet Sauvignon* – 169:176, just like in *Tămâioasă Românească*). The 176bp allele is frequentlyt encountered in Moldavian collection cultivars: *Muscat Timpuriu de București* (169:176), *Timpuriu de Cluj* (176:176), and *Apiren extratimpuriu* (176:176) (Table 2 and Table 3). Interesting, although *Muscat Timpuriu de București* and *Timpuriu de Cluj* have Romanian origin, and *Apiren extratimpuriu* is a newly created cultivar, its parents having probably a Caucazian origin, the allelic variant of 176bp has a good penetrance and has been stabilized in the Moldavian grapevine gene pool.

The **VVMD6** locus was identified on the linkage group corresponding to chromosome 7 of the *Vitis vinifera* complement (Lowe et al., 2006; Riaz et al., 2004; Troggio et al., 2007; Vezzulli et al., 2008). At this SSR locus, in the 14 analysed cultivars, 7 allelic variants for Romanian- and also for Moldavian cultivars (Table 4 and Table 5) have been detected. The most frequent allelic variant in Romanian cultivars is that of 200bp (approximately 36%), and in Moldavian cultivars – that of 206bp (approximately 36%). Some of the alleles, shorter than the variants cited in the literature for this locus (Table 1), appear in both stock collections: *Fetească neagră* (R) – 186:186; *Fetească regală* (R) – 187:197; *Sauvignon* (R) – 191:200; *Cabernet Sauvignon* (M) – 173:187 (Table 2 and Table 3). The size and frequency of the alleles differ between the two stock collections.

The **VVMD7** locus was identified on the linkage group corresponding to chromosome 7 of the *Vitis vinifera* complement (Adam-Blondon et al., 2004; Lowe et al., 2006; Riaz et al., 2004; Troggio et al., 2007; Vezzulli et al., 2008). This SSR locus presents a high allelic variability, being selected by the *Vitis* Working Group among the recommended SSR markers. In Romanian cultivars, 8 allelic variants have been found, most frequent being those of 241bp and 248bp (approximately 21,5%) (Table 4 and Table 5). For Moldavian cultivars, only 4 different alleles have been found in the 7 analysed cultivars, with the 240bp allele presenting highest frequency (approximately 57%). 3 of the Moldavian cultivars are homozygous- and 2 are heterozygous for this allele (Table 2 and Table 3). In our previous researches (Ghețea et al., 2010, b; unpublished data) we have found a greater variability and also a relative high number of repetitions in sequenced amplicons of different allelic variants, which indicate that the microsatellite locus is placed in an non-coding region with high mutational rates (Oliveira et al., 2006).

ZAG7	FA (%)	ZAG12	FA (%)	ZAG15	FA (%)	ZAG62	FA (%)	ZAG79	FA (%)
155	**42,9**	140	7,1	162	21,4	188	7,1	228	8,3
157	**42,9**	147	7,1	**164**	**50,0**	190	14,3	230	16,8
158	14,2	**148**	**43,0**	172	14,3	194	14,3	235	16,8
		150	28,5	174	14,3	196	28,6	**246**	**24,9**
		170	14,3			**198**	**35,7**	247	8,3
								248	8,3
								252	8,3
								254	8,3

VVS1	FA (%)	VVS2	FA (%)	VVS4	FA (%)	VVMD6	FA (%)	VVMD7	FA (%)
176	21,4	**128**	**50,0**	167	14,3	186	14,3	235	14,3
177	14,3	130	7,1	**168**	**42,9**	187	7,1	240	7,1
178	**35,8**	135	7,1	169	21,4	191	7,1	**241**	**21,5**
180	14,3	139	14,3	176	21,4	197	7,1	**248**	**21,5**
183	7,1	148	21,5			199	14,3	250	7,1
186	7,1					**200**	**35,8**	254	7,1
						207	14,3	256	7,1
								258	14,3

VVMD 14	FA (%)	VVMD 17	FA (%)	VVMD 21	FA (%)	VVMD 27	FA (%)	VVMD 36	FA (%)
216	**71,5**	209	7,1	239	7,14	172	7,2	239	7,1
218	7,1	210	14,3	243	14,3	176	14,3	**249**	**28,6**
222	7,1	218	14,3	**244**	**28,6**	182	21,4	**259**	**28,6**
228	14,3	**219**	**50,0**	245	21,4	186	21,4	261	14,3
		220	14,3	246	7,14	**191**	**35,7**	267	14,3
				253	7,14			271	7,1
				255	7,14				
				263	7,14				

Table 4. Frequencies of the allelic variants found at each analysed microsatellite locus, in Romanian stock collection cultivars. Bolded numbers represent the most frequent allele/alleles.

The **VVMD14** locus was identified on the linkage group corresponding to chromosome 5 of the *Vitis vinifera* complement (Riaz et al., 2004). For the 14 analysed cultivars, 4 allelic variants (with the dominant allele of 216bp – approximately 71,5%) have been found in Romanian cultivars, and 7 allelic variants – in Moldavian ones (with the allele of 217bp having the highest frequency - 33%) (Table 4 and Table 5). In both cultivar groups appear shorter allelic variants (216bp, 217bp, 219bp), also in an homozygotic- and heterozygotic condition (Table 2 and Table 3). These shorter alleles could represent a particular feature for the grapevine gene pool from this East-European region.

The **VVMD17** locus was located on the linkage group corresponding to chromosome 18 of the *Vitis vinifera* complement (Adam-Blondon et al., 2004; Troggio et al., 2007; Vezzulli et al., 2008). A moderate variability was revealed by this microsatellite locus: 5 allelic variants in Romanian cultivars (the most frequent alelic variant is that of 219bp – 50%), and 6 variants in Moldavian cultivars (the allele of 220bp has the highest frequency – 35,5%) (Table 4 and Table 5). The relatively low number of allelic variants, and also short repetitive sequences (Gheţea et al., 2010, b) found at this SSR locus, indicate that VVMD17 is probably placed in a coding region of the genome, where the mutational rates are low (Oliveira et al., 2006). A shorter allelic variant than the size interval cited in the literature (Table 1), of 209/210bp, in a heterozygous condition, appears in 3 Romanian cultivars: *Fetească albă*, *Fetească regală* and *Fetească neagră*. One can say that all these 3 cultivars have the same genetic constitution at this locus – 209/210:219 (Table 2). In Moldavian cultivars, all the allelic variants fit in the allelic size interval, having a homozygous either heterozygous constitution (Table 2 and Table 3).

The **VVMD21** locus is placed on the linkage group corresponding to chromosome 6 of the *Vitis vinifera* complement (Adam-Blondon et al., 2004; Riaz et al., 2004; Lowe et al., 2006). Most of the analysed cultivars presented a heterozygous status for this locus, where we found 8 allelic variants in Romanian cultivars, and 7 different alleles in Moldavian cultivars. In Romanian group, the most frequent allele is that of 244bp (approximately 29%), while in Moldavian group, the highest frequency presented the 245bp allele (approximately 25%) (Table 4 and Table 5). Excepting the short allele of 239bp (found in *Fetească neagră* from Romania), all the other allelic variants fit in the known size interval (Table 1, Table 2 and Table 3).

The **VVMD27** locus is placed on the linkage group corresponding to chromosome 5 of the *Vitis vinifera* complement (Lowe K.M et al., 2006; Riaz et al., 2004; Troggio et al., 2007; Vezzulli et al., 2008). This microsatellite region is among the selected loci as SSR marker for grapevine genetic profiling, due to its high allelic variability revealed among different cultivars. The heterozygous constitution of most of the analysed cultivars indicate that this locus is a mutational dynamic one, placed most probably, in a non-coding region of the grapevine genome. In the 14 cultivars, we found 5 different allelic variants in Romanian group (with the most frequent allele – that of 191bp – approximately 36%), and 8 allelic variants in Moldavian group (here, two different alleles, one - the same as in Romanian cultivars – 182bp and 191bp – presented the highest frequency – approximately 21%) (Table 4 and Table 5). Excepting the *Cabernet Sauvignon* cultivar from Romania (which has an allele of 172bp – at the limit of the size interval), all other allelic variants are in the cited size range (Table 1, Table 2 and Table 3).

ZAG7	FA (%)	ZAG12	FA (%)	ZAG15	FA (%)	ZAG62	FA (%)	ZAG79	FA (%)
102	14,3	**146**	28,6	163	28,6	189	7,1	**237**	25,1
103	14,3	147	14,3	**164**	57,2	190	50,2	238	8,3
107	14,3	153	14,3	165	7,1	195	7,1	246	8,3
156	28,6	155	7,15	192	7,1	196	14,3	248	16,7
157	14,3	156	7,15			200	7,1	250	16,7
158	14,3	165	14,3			205	7,1	253	8,3
		166	14,3			206	7,1	254	8,3
								258	8,3

VVS1	FA (%)	VVS2	FA (%)	VVS4	FA (%)	VVMD6	FA (%)	VVMD7	FA (%)
170	7,1	126	14,3	169	24,9	173	7,2	**240**	57,1
178	57,3	128	7,1	**176**	58,1	187	7,2	244	14,3
179	14,3	129	14,3	177	17,0	198	7,2	250	14,3
185	7,1	131	7,1			200	7,2	252	14,3
187	7,1	139	7,1			**206**	35,5		
188	7,1	**141**	21,5			207	28,5		
		148	21,5			212	7,2		
		153	7,1						

VVMD 14	FA (%)	VVMD 17	FA (%)	VVMD 21	FA (%)	VVMD 27	FA (%)	VVMD 36	FA (%)
217	33,2	217	7,3	243	16,6	176	7,2	103	12,5
218	8,4	218	7,3	**245**	25,0	177	7,2	234	6,25
219	8,4	219	28,4	246	16,6	178	14,3	236	12,5
225	8,4	**220**	35,5	247	8,4	179	14,3	237	6,25
228	8,4	221	14,2	252	8,4	180	7,2	240	12,5
232	16,6	231	7,3	253	16,6	**182**	21,3	259	12,5
236	16,6			255	8,4	186	7,2	260	12,5
						191	21,3	262	6,25
								264	6,25
								266	6,25
								272	6,25

Table 5. Frequencies of the allelic variants found at each analysed microsatellite locus, in Moldavian stock collection cultivars. Bolded numbers represent the most frequent allele/alleles.

The **VVMD36** locus was located on the linkage group corresponding to chromosome 3 of the *Vitis vinifera* complement (Adam-Blondon et al., 2004; Riaz et al., 2004; Troggio et al., 2007; Vezzulli et al., 2008). This is a very interesting microsatellite locus, which revealed particular genetic constitutions, especially for Moldavian cultivars. There is a high genetic variability at this locus, a real mutational „hot spot", generating many allelic variants and also revealing multiallelic profiles in some cultivars. In our researches, we found an important number of Moldavian cultivars having more than 2 different alleles at VVMD36 locus. Triallelic (Fig.2) and tetraallelic constitutions (Fig.3) are common in Moldavian grapevine gene pool. Here, we present 2 cultivars: *Cabernet Sauvignon* (103:237:262), and *Apiren extratimpuriu* (103:234:236) (Table 2 and Table 3).

Another phenomenon is also obvious at this SSR locus: a significant allele shortening tendency, an important number of base pairs being deleted. The allele of 103bp is more than a half smaller than the inferior limit of the allelic size interval known for this SSR locus (Table 1). Based on these results, we have presumed that in some microsatellite loci, gene duplication processes, followed by deletion or repeated deletions, take place.

These multiallelic profiles are interpreted by Moncada (Moncada et al., 2006) as a result of either a gene duplication phenomenon, or chimeras produced by mutation in meristematic layers L1 and L2. Such events could be considered as local evolutionary mechanisms that give, in time, particular features of a certain gene pool. Certainly, multiallelic profiles at SSR locus represent a characteristic feature for Moldavian grapevine germplasm collection.

At this locus, considering the 14 cultivars, we found 6 allelic variants in Romanian group (with 2 most frequent alleles, those of 249bp and 259bp – approximately 29% each), and 11 allelic variants in Moldavian cultivars (Table 4 and Table 5).

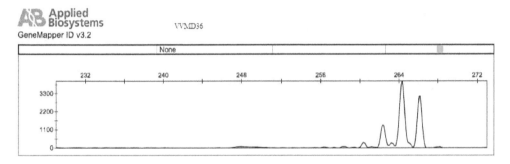

Fig. 2. Capillary electrophoresis showing a trialellic profile (262:264:266) at VVMD36 locus, for a cultivar from Moldavian stock collection (data not published)

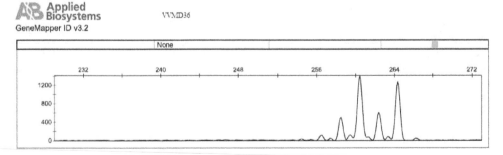

Fig. 3. Capillary electrophoresis showing a tetraalellic profile (258:260:262:264) at VVMD36 locus, for another cultivar from Moldavian stock collection (data not published)

Considering the two non-autochthonous cultivars – *Cabernet Sauvignon* and *Sauvignon* – we
have compared genetic profiles obtained by us, for these two famous and geographical
highly dispersed cultivars, to other citations from the literature. Table 6 and Table 7
summarize the results. Although we do not have the data for all the SSR loci we have
considered in these tables, the comparison sustains the idea that discrete, molecular
evolutive mechanisms are developing in every regional gene pool, giving, in time, unique
particularities which make it an important and valuable genetic reservoir.

		ZAG62	ZAG79	VVS2	VVMD6	VVMD7
Romania	*Cabernet Sauvignon*	190 196	228 247	135 148	200	241
Republic of Moldova		190 205	248 258	128 141	173 187	244
Martin et al., 2003		187 193	245	136 150	-	237
Moncada et al., 2006		-	-	-	204 205	236

Table 6. Allelic variants (in base pair number) in 5 SSR loci, determined in *Cabernet
Sauvignon* cultivar from Romanian- and Moldavian stock collections, compared to other
citations for the same cultivar

		ZAG62	ZAG79	VVS2	VVMD6	VVMD7	VVMD27
Romania	*Sauvignon*	190 196	235	128 148	191 200	241 258	182 191
Santiago et al., 2007		187 193	243 249	150	-	241 255	185
Martin et al., 2003		187 193	243 249	-	-	241 255	-
Moncada et al., 2006		-	-	-	197 205 209	236 254	-

Table 7. Allelic variants (in base pair number) in 6 SSR loci, determined in *Sauvignon* cultivar
from Romanian stock collection, compared to other citations for the same cultivar

6. Conclusion

The 15 microsatellite loci we have used, proved to be reliable molecular markers in assessing
the genetic profile of the studied grapevine cultivars. A high number of microsatellite
markers allow for a rigorous genetic characterization, a fine discrimination between

different cultivars and also pedigree analyses, in order to assess the origins, oldness and phylogenetic relationships between grapevine cultivars.

Some of these SSR loci showed high variability: for the cultivars from Romanian stock collection, the most variable loci have been VrZAG79, VVMD6, VVMD7, and VVMD21; for the cultivars from Moldavian stock collection, high numbers of allelic variants have been found in VrZAG12, VrZAG62, VrZAG79, VVS2, VVMD6, VVMD14, VVMD21, VVMD27 and VVMD36.

Besides a great allelic variability evidenced in these microsatellite loci, a detailed analysis of the genetic constitution of the grapevine cultivars, at each SSR locus, revealed also allelic variants well conserved inside a certain cultivar group (Romanian or Moldavian); these alleles are specific for that germplasm collection. There are also allelic forms conserved in both groups, proving, in some cases, the existence of a phylogenetic relationship between cultivars, or in other cases, being the consequence of the close vicinity and of a multimillenary tradition in grapevine cultivation, of Romania and Republic of Moldova.

Lower variability in some SSR loci (like VrZAG7, VVS4, VVMD17) can be explained by the fact that, very probably, these microsatellite regions are placed in coding regions of the genome, where the mutational rate is low. The conservation of certain allelic variants in these loci is stimulated also by the self-pollination or by repeated artificial selection and vegetative multiplication, if these SSR loci are associated with genes coding for economically important traits.

For Moldavian germplasm, the VVMD36 microsatellite locus proved to be very informative, revealing interesting multiallelic profiles. Significantly shorter allelic variants, and also triallelic loci, have been detected. The observation proves that these shorter alleles and the existence of 3 (or even 4) instead of 2 allelic forms at VVMD36 locus, are specific well conserved genetic features for the Moldavian grapevine gene pool. These results sustain the hypothesis of the existence of an evolutive event at this microsatellite locus: a gene duplication phenomenon followed by a deletion or repeated deletion events could explain this situation.

High genetic variability in a species is a sign of its health, and for an economically important species, it represents a valuable source for agricultural purposes. The final conclusion is that both – Romanian and Moldavian – germplasm collections are characterized by a high diversity, with specific allelic combinations which give unique and valuable features that must be known and inventored, enabling their long-term conservation and use in breeding programs.

7. Acknowledgment

The research was supported by the Romanian Education and Research Ministry, through the Second National Plan for Research, Development and Innovation, grant no. 51-037 (2007-2010), grant no. 1072 (2009-2011), and by the international grant INTAS 05-104-7654 (2006-2008).

8. References

Adam-Blondon, A.F., Roux, C., Claux, D., Butterlin, G., Merdinoglu, D. & This P. (2004). Mapping 245 SSR markers on the *Vitis vinifera* genome: a tool for grape genetics. *Theor. Appl. Genet.*, Vol.109, pp.1017-1027, ISSN 0040-5752.

Ausubel, F., Brent, R., Kingston, R., Moore, D., Seidman, J., Smith, J. & Struhl, K. (1990). *Current Protocols in Molecular Biology*, Vol.1, John Wiley and Sons, New York, USA.

Bowers, J.E., Dangl, G.S., Vignani, R. & Meredith, C.P. (1996). Isolation and characterization of new polymorphic simple sequence repeat loci in grape (*Vitis vinifera* L.). *Genome*, Vol.39 No.4, pp.628-633, ISSN 0831-2796.

Bowers, J.E, Dangl, GS, Meredith, C.P. (1999). Development and characterization of additional microsatellite DNA markers for grape. *Am. J. Enol. Vitic.*, Vol.50, No.3, pp.243-246, ISSN 002-9254.

Dobrei, A., Rotaru, L. & Mustea, M. (2005). *Cultura viței de vie*, Solness (ed.) ISBN 978-973-729-120-2, Timişoara, Romania.

Gheţea, L.G., Motoc, R.M., Popescu, C.F., Barbacar, N., Iancu, D., Constantinescu, C. & Barbarii, L.E. (2010, a). Genetic profiling of nine grapevine cultivars from Romania, based on SSR markers. *Romanian Biotechnological Letters*, Vol.15, No.1, Supplement (2010), pp.116-124, ISSN 1224-5984.

Gheţea, L.G., Motoc, R.M., Popescu, C.F., Barbacar, N., Iancu, D., Constantinescu, C. & Barbarii, L.E. (2010, b). Genetic variability revealed by sequencing analysis at two microsatellitic loci, in some grapevine cultivars from Romania and Republic of Moldavia. *Romanian Biotechnological Letters*, Vol.15, No.2, Supplement (2010), pp.116-124, ISSN 1224-5984.

Lowe, K.M. & Walker M.A. (2006). Genetic linkage map of the interspecific grape rootstock cross Ramsey (*Vitis champinii*) x Riparia Gloire (*Vitis riparia*). *Theor. Appl. Genet.*, Vol.112, Springer-Verlag, Online ISSN 1432-2242, DOI:10.1007s00122-006-0264-8 (April, 2006).

Martin, J.P., Borrego, J., Cabello, F. & Ortiz, J.M. (2003). Characterization of Spanish grapevine cultivar diversity using sequence-tagged microsatellite site markers. *Genome*, Vol.46, No.1, pp.10-18, ISSN 0831-2796.

Meredith, C.P. (2001). Grapevine Genetics: Probing the past and facing the future. *Agriculturae Conspectus Scientificus*, Vol.66, No.1, pp.21-25, ISSN 1331-7768.

Moncada, X., Pelsy, F., Merdinoglu, D. & Hinrichsen P. (2006). Genetic diversity and geographical dispersal in grapevine clones revealed by microsatellite markers. *Genome*, Vol. 49, pp.1459-1472, Online ISSN 1480-3321, DOI:10.1139/G06-102. Published at http://genome.nrc.ca (January, 2007).

Oliveira, E.J., Pádua, J.G., Zucchi, M.I., Vencovsky, R. & Vieira M.L.C. (2006). Origin, evolution and genome distribution of microsatellites. *Genetics and Molecular Biology*, Vol.29, No.2, pp.294-307, ISSN 1415-4757.

Riaz, S., Dangl, G.S., Edwards, K.J. & Meredith, C.P. (2004). A microsatellite marker based framework linkage map of *Vitis vinifera* L. *Theor. Appl. Genet.*, Vol.108, pp.864-872, Springer-Verlag, Online ISSN 1432-2242, DOI:10.1007/s00122-003-1488-5 (November 2003).

Santiago, J.L., Boso, S., Gago, P., Alonso-Villaverde, V. & Martínez, M.C. (2007). Molecular and ampelographic characterization of *Vitis vinifera* L. *'Albăriño'*, *'Savagnin Blanc'* and *'Caíño Blanco'* shows that they are different cultivars. *Spanish Journal of Agricultural Research*, Vol.5, No.3, pp.333-340 ISSN 1695-971-X.

Savin, Gh. (2010). Conservation and using of grapevine genetic resources in the Republic of Moldova for pre-breeding stage. *Romanian Biotechnological Letters*, Vol.15, No.2, Supplement (2010), pp.113-116, ISSN 1224-5984.

Sefc, K.M., Regner, F., Turetschek, E., Glössl, J. & Steinkellner, H. (1999). Identification of microsatellite sequences in *Vitis riparia* and their applicability for genotyping of different *Vitis* species. *Genome*, Vol.42, pp.367-373, ISSN 0831-2796.

Sefc, K.M., Lopes, M.S., Lefort, F., Botta, R., Roubelakis-Angelakis, K.A., Ibáñez, J., Peji´c, I., Wagner, H.W., Glössl, J. & Steinkellner, H. (2000). Microsatellite variability in grapevine cultivars from different European regions and evaluation of assignment testing to assess the geographic origin of cultivars. *Theor. Appl. Genet.*, Vol.100, pp.498-505, ISSN 0040-5752.

Sefc, K.M., Lefort, F., Grando, M.S., Scott, K.D., Steinkellner, H. & Thomas, M.R. (2001). Microsatellite markers for grapevine: a state of the art. In: *Molecular Biology and Biotechnology of Grapevine*. K.A. Roubelakis-Angelakis (ed.), pp.1-30, Kluwer Academic Publishers, ISBN 0-7923-6949-1,The Netherlands.

Scott, K.D., Eggler, P., Seaton, G., Rosetto, E.M., Ablett, E.M., Lee, L.S. & Henry, R.J. (2000). Analysis of SSRs derived from grape ESTs. *Theor. Appl. Genet.*, Vol.100, pp.723-726, ISSN 0040-5752.

Thomas, M.R. & Scott, N.S. (1993). Microsatellite repeats in grapevine reveal DNA polymorphisms when analysed as sequence-tagged sites (STSs). *Theor. Appl. Genet.*, Vol.86, No.8, pp.985-990, ISSN 0040-5752.

Thomas, M.R., Scott, N.S., Botta, R. & Kijas, M.H. (1998). Sequence-tagged site markers in grapevine and citrus. *Journal of the Japanese Society for Horticultural Science*, Vol.67, pp.1189- 1192, ISSN 0013-7626.

Troggio, M., Malacarne, G., Coppola, G., Segala, C., Cartwright, D.A., Pindo, M., Stefanini, M., Mank, R., Moroldo, M., Morgante, M., Grando, S.M. & Velasco, R. (2007). A dense single-nucleotide polymorphism-based genetic linkage map of grapevine (*Vitis vinifera* L.) anchoring *Pinot Noir* bacterial artificial chromosome contigs. *Genetics*, Vol.176, pp.2637-2650, ISSN 0016-6731.

Vallejos, C.E., Sakiyama, N.S. & Chase, C.D. (1992). A molecular-based linkage map of *Phaseolus vulgaris* L. *Genetics*, Vol.131, pp. 733-740, ISSN 0016-6731.

Vezzulli, S., Troggio, M., Coppola, G., Jermakow, A., Cartwright, D., Zharkikh, A., Stefanini, M., Grando, S.M., Viola, R., Adam-Blondon, A.F., Thomas, M., This, P. & Velasco, R. (2008). A reference integrated map for cultivated grapevine (*Vitis vinifera* L.) from three crosses, based on 283 SSR and 501 SNP-based markers. *Theor. Appl. Genet.*, pp.1-13, Springer-Verlag, Online ISSN 1432-2242, DOI:10.1007/s00122-008-0794-3.

Vitis Working Group: http://www.ecpgr.cgiar.org/workgroups/vitis/

Part 3

Protected Horticulture

4

Total Growth of Tomato Hybrids Under Greenhouse Conditions

Humberto Rodriguez-Fuentes[1], Juan Antonio Vidales-Contreras[1],
Alejandro Isabel Luna-Maldonado[1] and Juan Carlos Rodriguez-Ortiz[2]
[1]*Department of Agricultural and Food Engineering, Faculty of Agriculture,
Autonomous University of Nuevo Leon, Escobedo, Nuevo Leon,*
[2]*Faculty of Agriculture, Autonomous University of San Luis Potosi, San Luis Potosi,
Mexico*

1. Introduction

Often in intensive production of tomato, the fertilization is applied by the farmers without consider the suitable doses in order to cover nutritional requirements according to crop physiological stages. Thus, appropriate crop management is a strategic demand to maintain or increase tomato production. In spite of many researchers conducted experiments in this subject and data is available about physiological stage requirements for plant nutrition, only few studies have been focused to nutritional parameters. The crop growth curves and nutrient uptake for tomato may determine uptake rate for a particular nutrient eluding possible deficiencies and superfluous fertilizer consumption. The daily rates of nutrient uptake are depending on crop and wheater (Scaife and Bar-Yosef, 1995; Honorato *et al.*, 1993; Magnificent *et al.*, 1979; Miller *et al*, 1979); however, crop requirements and opportune fertilizer applications, are little known in many of fresh consumption crops. In Mexico, vegetable production is located at desert areas in the north and middle of the country where water shortages have constraints with impact on of water demands the crops of tomato, pepper and cucumber. Thus, the surface for crop production in greenhouses has increased from 350 ha in 1997 (Steta, 1999) to about 5000 ha in 2006 (Fonseca, 2006), because the increasing demand for quality products and the risk of losses on field for crop production.

Imas (1999) found that nutrient uptake and fertilization recommendations are conditions depending of crops. For example, tomato crops under hydroponic greenhouse environments have averaged 200 t ha^{-1} which is significantly higher than the 60-80 t ha^{-1} yield, typically observed in an open field. In contrast to tomato crops grown in an open field, nutrient uptake in greenhouse environment can be duplicated or triplicated. In practical terms, crop growth cycle is divided according to physiological stages thence different concentrations or amounts of nutrients have to be applied according to recommendations given by Department of Agriculture. In the tomato production are considered four physiological stages: establishment-flowering, flowering-fruit set, ripening of tomato fruit (the first-crop harvest and last harvest on the crop). In each stage concentrations of nitrogen (N) and phosphorus (K) are increasing while nitrogen-phosphorous are decreasing because potassium is uptaken in large quantities during the reproductive stage of the crop (Zaidan and Avidan, 1997). In order to determine nutrient uptake more accurately, crop growth can be scheduled by chronological periods of sampling and analysis nearest one another.

The nutrient demand is the maximum amount of nutrient that a crop needs in order to meet their metabolic growth demand and development and that is calculated to maximize production goals and domestic demand price (nutritional optimal concentration of total biomass (air and / or root) at harvest time); however this criteria is not yet thought by the farmers. An appropriate method for such calculation is to use mass balance concept in a hydroponic system. In this way, nutrimental control is more efficient in the nutritive solution and its effect can be seen rapidly in the plants (Steiner, 1961). The method before mentioned is based on the composition of plant dry matter which consists of 16 essential nutrients; however, only thirteen are directly uptaken by plants from soil. Therefore, if the amount of total dry matter production (root + aerial part) during the cycle of growth and development as well as nutrimental concentration in each physiological stage is determined, the amount of nutrients that the plant absorbs can be estimated. With this analysis it is possible to establish a program of daily/weekly fertilization for crop production. This would result in a substantial economic savings on fertilizer costs and also in a decrease of negative environmental impact for its inappropriate applications.

2. Tomato hybrids analysis

The field study was conducted on the farm "El Cuento" located in the town of Marin, Nuevo León, Mexico (latitude N 25° 53' and longitude W 100° 02' and 400 m above sea level), where two greenhouses were used (Figure.1), first one of these greenhouses was a tunnel type of 900 m^2 (75 m x 12 m and 7 m high) and second one was multi-Korean tunnel type of 1300 m^2 (45 m x 30 m). It was used an open hydroponics system during the tests of two indeterminate hybrid tomatoes beef type.

The seeds were sown in a mixture of peat moss and perlite (1:1 v/v) in containers of 200 cavities. The transplanting was realized at 40 days after sowing. Plants were transplanted inside of 2 gallons polythene bags (white outer and black inner, Fig.2) with a mixture of perlite and peat moss (1:1, v/v). In the Korean type greenhouse it was settled 50% of plants of the hybrids Cayman (1700) and Charleston (1700). In the tunnel type greenhouse it was established 2220 plants of the hybrid Charleston, in both cases it was used a density of 2.5 plants m^{-2}. The hydroponic irrigation system and nutrition of the plants were conducted with emitters calibrated to 4 min L^{-1}. It was drained of 10 to 15% of the hydroponic solution applied. The nutrimental solution used was suggested by Rodríguez et al. (2006).

Plant sampling and were conducted according to tomato crops and greenhouse type (Fig 3). On plant sampling, three plants, visually uniform, were removed from the hydroponic system (substrate) every 15 days after plant plantation total dry matter was determined considering root + leaves + stems + flowers and fruits (if they were in the plant) To obtain the biomass model in both production and the nutrient uptake, roots, stems, leaves, and fruits were included. The samples were analyzed in the Laboratory of Soil, Water and Plant Analysis at the Agronomy School of Autonomous University of Nuevo Leon, where plant specimens were washed with deionizer water to obtain their dried weights. In addition to this parameter, dry matter, nitrogen, phosphorus, potassium, calcium and magnesium were determined per plant sample.

Fig. 1. Tunnel-type greenhouse used for experimentation located in Marin, Nuevo Leon, Mexico.

Fig. 2. Vinyl bag used for growing tomato in a hydroponics system.

Fig. 3. Tomato plants inside of the experimental greenhouse.

To determine the dry weight at constant weight, samples were milled and then placed in a forced convection oven (Riossa model F-62, Mexico) at 70° to 80°C. Samples were screened through a 50 μm mesh. Determination of total nitrogen was done by the Kjeldhal method (Rodríguez and Rodríguez, 2011). A wet digestion microwave (MARSX, EMC Corporation, North Carolina, USA) was used in order to analyze P, K, Ca, and Mg. Three dried samples of 0.5 ± 0.001g were placed in a digestion flask (Teflon PFA vessels of a capacity of 120 mL) and then 5 mL of HNO_3 were added. The instrument was programmed with a ramp of 15 min to reach a temperature of 180 degrees Celsius and a pressure of 300 PSI and keeping these conditions for 5 minutes. Finally, to the samples were left to cool for 15 minutes.

The dry matter accumulated values in the hybrid tomato were analyzed with the DM Sigma Plot 10.0 Program and Microsoft Excel Office 2003 software and it was found that the best fitted linear regression model. To calculate the equations were considered the average values of both hybrids.

3. Flowering period

The flowering period occurred after 20 days after plant plantation, fruits set showed up 10 days later and until 55 days after transplanting. During the beginning of flowering there was a low accumulation of dry matter. On the fruit set and harvest time of fruit an accumulation of intense biomass was observed (Figures. 4 and 5) until the establishment of the harvest. In last stage of crop growth the dry matter is decreasing as well as the nutrient uptake. Bugarin *et al.* (2002), coincided in their research due in same crop growth stage any nutritional deficit decreases production.

Fig. 4. Tomato plants (variety Charleston) in breaking stage.

4. Accumulation of dry matter production

Accumulation of dry matter production had a linear behavior up to 96-day initial crop period. Thereafter, an erratic behavior was observed probably as a consequence of the conditions of high temperatures and low relative humidity occurred in those greenhouse designs. This condition caused a severe reduction in dry matter accumulation, low pollination of fruit and fruit set. Sampling finished after 170 days since transplanting stage; however, plant health recovery did not occurred, reducing drastically fruit production to about one third of the best period, from 2.5 to 0.8 kg m² after 125 days since transplanting. The same reduction is normal but it occurs gradually as the harvest progresses and the number of clusters.

Observed data sets were analyzed for dry matter production during 96 days after transplanting. Three models of linear fitting (Table 1 and Figures 6, 7, 8) with R^2 values of 0.96, 0.99 and 0.91 for hybrids: Cayman (CAI), Charleston greenhouse Korean (CHK) and Charleston in greenhouse Tunnel (CHT), respectively, were obtained. The R^2 value for the average of both hybrids was 0.96 (Table 1, Figs. 6 and 7) to predict possible accumulation of DM with the models obtained for each hybrid, is to believe the 170 days scheduled time to remove crop residues and begin the next crop season.

Fig. 5. Pruning of unnecessary tomato shoots.

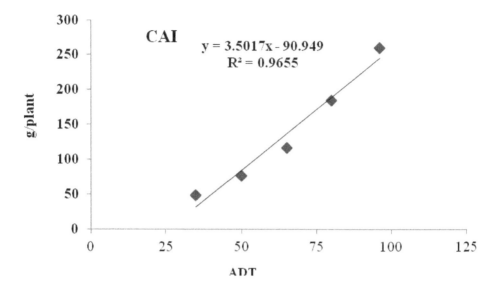

Fig. 6. Dry matter and total accumulated average in CAI: Caiman; DM A: Dry Matter average; ADT: After day trasplanting.

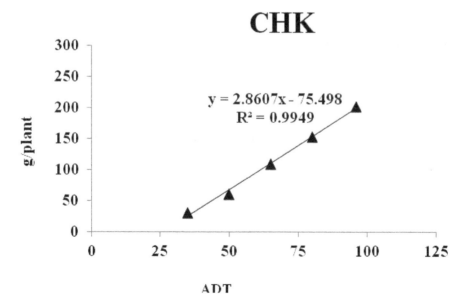

Fig. 7. Dry matter and total accumulated average in CHK: Charleston Korean; DM A: Dry Matter average; ADT: After day trasplanting.

CHT

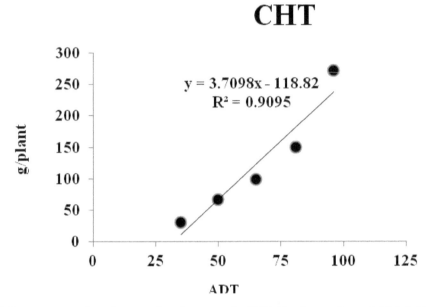

$$y = 3.7098x - 118.82$$
$$R^2 = 0.9095$$

Fig. 8. Dry matter and total accumulated average in CHT: Charleston Tunnel; DM A: Dry Matter average; ADT: After day trasplanting.

ADT	DMA (g/plant)			
	CAI	CHK	CHT	Average
35	48.69	30.77	30.77	36.74
50	76.81	60.47	66.66	67.98
65	116.16	108.98	98.76	107.96
80	184.80	153.06	150.00	162.62
96	260.32	201.80	272.77	244.96
R^2	0.9655	0.9949	0.9095	0.9672
Rate of Increase of MS g/day/plant	3.5017	2.8607	3.7098	3.3699

Table 1. Dry matter and total accumulated average in two hybrid tomatoes. CAI: Caiman; CHK: Charleston grown in Korean greenhouse; CHT: Charleston in tunnel; DMA: Dry matter average (g/plant). ADT: after day transplanting.

Figure 9 shows the estimated model with the average values in the three systems evaluated, if there were some mathematical models of nutrient removal, we could estimate the dose for each nutrient to apply, once it has the most of the nutrients, it is possible calculate a fertilization program for growing tomatoes, a further advantage of this procedure is that it could adjust fertilization with irrigation, if available, a pressurized irrigation system.

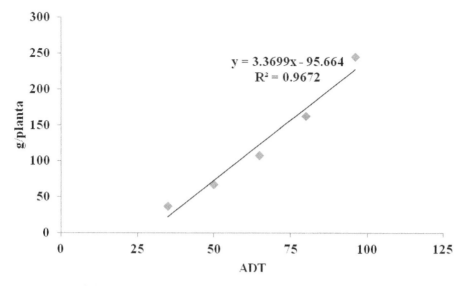

Fig. 9. Linear model obtained from total dry matter considering both hybrids.

5. Conclusion

Although the stage of harvest of tomato is not completed successfully because of environmental factors, It was possible to estimate dry matter production, the rate of increase in dry matter was higher and similar for CAI and CHT, but was lower for CHK. The average yield of the two hybrids was considered adequate to estimate dry matter production in tomato crop in this geographical area. Once we determined the removal of nutrients such as nitrogen, phosphorus, potassium, among others, and generate models of extraction of these, it can be calculated the amount of each nutrient applied and thus may establish a program of fertilization to ensure a sustainable production system; such program may be adjusted by several cycles of this same culture.

6. Acknowledgment

The authors would like to thank to M Sc. Gerardo Jiménez García[†] his contribution to accomplish this research.

7. References

Bugarin, M. R.; Galvis, A.; Sanchez, P. y García, D. (2002). *Acumulación diaria de materia seca y de potasio en la biomasa aérea total de tomate*. Terra Latinoamericana. Vol.20, No.4, pp. 401-409, ISSN 0187-5779

De Koning, A.N.M. (1989). *Development and growth of a commercially grown tomato crop*. Acta Hort. Vol.260, pp. 267-273, ISSN 0567-7572

Galvis, S. A. (1998). *Diagnóstico y simulación del suministro edáfico para cultivos anuales*. Tesis Doctoral en Ciencias. Colegio de Postgraduados. Montecillo, Mexico.

Gary, C.; Jones, J.W. & Tchamitchian, M. (1998). *Crop modelling in horticulture: State of the art.* Sci. Hort. Vol.74, pp. 3-20, ISSN 03044238

Heuvelink, E. & Marcelis, L.F.M. (1989). *Dry matter distribution in tomato and cucumber.* Acta Hort. Vol.260, pp. 149-157. ISSN 0567-7572

Honorato, R.; Gurovich, L. y Piña R. (1993). *Ritmo de absorción de N, P y K en pepino de semilla.* Cien. Inv. Agr. Vol.20, pp. 169-172 ISSN 0718-3267

Imas, P. (1999). *Manejo de Nutrientes por Fertirriego en Sistemas Frutihortícolas.* XXII Congreso Argentino de Horticultura. Septiembre – Octubre, 1999. Argentina.

Magnífico, V.; Lattancio, V. & Sarli, G. (1989). *Growth and nutrient removal by broccoli raab.* J. Amer. Soc. Hort. Sci. 104 (2): 201 – 203. ISSN (print): 0394-6169

Miller, C.H.; Mccollum, R.E. & Claimon, S. (1979). *Relationships between growth of bell peppers (Capsicum annuum L.) and nutrient accumulation during ontogeny in field environments.* Jour. Amer. Soc. Hortic. Sci. Vol.104, pp. 852 – 857. Print ISSN: 0003-1062

Ortega B., R.; Correa-Benguria, M. y Olate M., E. (2005). *Determinación de las curvas de acumulación de nutrientes en tres cultivares de lilium spp. para flor de corte.* Agrociencia Vol.40, No.1, pp. 77-88, ISSN 1405-3195

Rodríguez, J. (1990). *La fertilización de los cultivos: Un método racional.* Facultad de Agronomía. Pontificia Universidad Católica de Chile. Santiago de Chile. 406p.

Rodríguez F., H. y Rodríguez Absi, J. (2011). *Métodos de Análisis de Suelos y Plantas. Criterios de Interpretación.* Editorial Trillas S.A. de C. V. (Segunda edición). México. ISBN 978-607-17-0593-8.

Rodríguez F., H.; Muñoz, S. y Alcorta G., E. (2006). *El tomate rojo. Sistema hidropónico.* Editorial Trillas S.A. de C. V. (Primera edición). México. ISBN 968-24-7606-2

Scaife, A. & Bar-Yosef, B. (1995). *Nutrient and fertilizer management in field grown vegetables.* IPI Bulletin No. 13. International Potash Institute, Basel, Switzerland).

Steiner, A.A. (1961). *A universal method for preparing nutrient solutions of certain desired composition.* Plant and Soil. Vo.15, pp. 134-154, (print version) ISSN: 0032-079X

Steta, M. (1999). *Status of the greenhouse industry in Mexico.* Acta Hort. Vol.481, pp. 735-738. ISSN 0567-7572

Sigmaplot 8.0. (2003). *Users Manual.* Exact Graphs for Exact Science. Systat Software Inc, USA.

Zaidan, O. & Avidan, A. (1997). *Greenhouses tomatoes in soilless culture.* Ministry of Agriculture, Extension Service, Vegetables and Field Service Departments, USA.

Part 4

Postharvest Physiology

Chemical Composition and Antioxidant Activity of Small Fruits

Pranas Viskelis, Ramune Bobinaite, Marina Rubinskiene,
Audrius Sasnauskas and Juozas Lanauskas
Institute of Horticulture, Lithuanian Research Centre for Agriculture and Forestry
Lithuania

1. Introduction

Small fruits contain significant levels of micronutrients and phytochemicals with important biological properties. Consumption of small fruits has been associated with diverse health benefits, such as prevention of heart disease, hypertension, certain forms of cancer and other degenerative or age-related diseases (Manach et al 2004, Santos-Buelga and Scalbert 2000, Hummer and Barney, 2002). These beneficial health effects of small berry fruits could mostly be due to their particularly high concentrations of natural antioxidants (Wang et al., 1996), including phenolic compounds, ascorbic acid and carotenoids. Because of the high contents and wide diversity of health-promoting substances in berries, these fruits are often referred to as natural functional products (Bravo, 1998; Joeph et al., 2000). The chemical composition of berry fruits has previously been shown to be affected by the environmental conditions under which these plants are grown. However, accumulating data suggest that the genotype has a profound impact on the concentration and qualitative composition of phytochemicals and other important constituents of berries. Studies have reported that standard cultivars of dark-fruited berries present a higher antioxidant content compared to vegetables or other foods (Wang et al., 1997). The influence of the genotype is of increasing interest, and several studies, particularly those addressing antioxidants, have been published on this topic (Connor et al., 2002, Lister et al., 2002).

Therefore, the aim of this study was to evaluate the quality parameters (the total amounts of phenolic compounds, anthocyanins and ascorbic acid as well as radical scavenging capacity) related to the fruits of *Sambucus, Aronia, Ribes, Hippophae rhamnoides* and *Rubus* cultivars.

2. Materials and methods

The amount of total phenolics in the fruit extracts was determined with the Folin-Ciocalteu reagent according to the method of Slinkard and Singleton (1977) using gallic acid as a standard. The reagent was prepared by diluting a stock solution (Sigma-Aldrich Chemie GmbH, Steinheim, Germany) with distilled water (1:10, v/v). Samples (1 ml in duplicate) were aliquoted into test cuvettes, and 5 ml of Folin-Ciocalteu's phenol reagent and 4 ml of Na_2CO_3 (7.5%) were added. The absorbance of all samples was measured at 765 nm using a Genesys10 UV/VIS spectrophotometer (Thermo Spectronic, Rochester, USA) after incubation at 20°C for 1 h. The results were expressed as milligrams of gallic acid equivalent (GAE) per 100 g of fresh weight. The anthocyanins were evaluated spectrophotometrically (Giusti, Wrolstad, 2001).

The radical scavenging capacity against stable DPPH* was determined spectrophotometrically (Brand-Williams et al., 1995; Viskelis et al., 2010). Methanolic fruit extracts (5 g of homogenised berries were extracted with 50 ml of methanol) were tested for their antioxidant activity using the DPPH*. A stable DPPH* radical ($C_{18}H_{12}N_5O_6$, 2,2-diphenyl-1-picrylhydrazyl, M = 394.32 g/mol) was purchased from Sigma Aldrich Chemical, Germany and analytical reagent grade methanol was used (Penta, Czech Republic). When DPPH* (2 ml) reacts with an antioxidant compound (50 µl) that can donate hydrogen, the DPPH* is reduced, resulting in a colour change from from deepviolet to light-yellow. This change was measured every 1 min at 515 nm for 30 min. Radical scavenging capacity (RSC) was calculated by the following formula:

$$RSC=[(AB - AS) / AB] \times 100\%$$

where AB is the absorption of the blank sample ($t=0$ min), and AS is the absorption of the tested sample.

Total soluble solids (TSS) were determined by a digital refractometer (ATAGO PR-32, Atago company, Japan). The dry matter (DM) content was determined by the air oven method after drying at 105°C in a Universal Oven ULE 500 (Memmert GmbH+Co. KG, Schwabach, Germany) to a constant weight. The titratable acidity (TA) was measured by titrating 10 g of pulp that had been homogenised with 100 ml distilled water. The initial pH of the sample was recorded before titration with 0.1 N NaOH to a final pH 8.2. The acidity was expressed as the percentage of citric acid equivalent to the quantity of NaOH used for the titration. The carotenoid content, expressed as β-carotene, was estimated from the extinction value $E_{1\ cm}\ ^{1\%}$ =2592 at 450 nm (Scot, 2001). The absorbance of the hexane solution was determined at 450 nm against a hexane blank in a Genesys10 UV/VIS spectrophotometer (Thermo Spectronic, Rochester, USA). The ascorbic acid content was measured by titration with 2,6-dichlorphenolindophenol sodium salt solution using chloroform for intensely coloured extracts (AOAC, 1990).

The pH of the fruits was measured using a pH meter. The free and total ellagic acid contents of raspberries and that of anthocyanins in black currants were quantified using a reversed-phase high-performance liquid chromatography technique (Koponen et al., 2007; Rubinskiene et al., 2005).

Color coordinates (L*, a*, b*) measurements were made with a portable spectrophotometer MiniScan XE Plus (Hunter Associates Laboratory, Inc., Reston, Virginia, USA). The spectrophotometer was set to measure total reflectance with illuminant C and a 10° observation angle. The parameters L*, a* and b* (lightness, red value and yellow value, respectively, on the CIEL*a*b* scale) were measured, converted into hue angle ($h° = \arctan(b*/a*)$) and chroma ($C = (a*^2+b*^2)^{1/2}$). The spectrophotometer was calibrated on a standard white tile (X = 81.3, Y = 86.2, Z = 92.7) before each series of measurements.

Statistical analyses were performed using analysis of variance (ANOVA).

3. Black currant (*Ribes nigrum* L.) fruits

The use of black currants (*Ribes nigrum* L.) as a domesticated crop is comparatively recent and has occurred only within the last 400-500 years. The modern-day commercial cultivars of black currants are significantly advanced compared to their wild progenitors, and a range of desirable attributes have been selected within the available germplasm by breeders in Europe and elsewhere. The major objectives of further breeding projects for black currants are to obtain the following characteristics: combined resistance to fungal diseases, reversion and gall mites, suitability of bush habit for mechanical harvesting, high productivity, self-fertility, spring frost resistance or avoidance, winter hardiness, adaptation to minimal pruning and suitability for juice production and other products (Keep, 1975; Brennan et. al., 1993; Viskelis et al., 2008). These objectives are most readily achieved by crosses between regional forms of *R. nigrum* L. and species of different sections of the *Eucoreosma* subgenus (Brennan, 1990). Black currants are widely used to make juices, wines, soft drinks and various preserved products. They are associated with important high-value horticultural industries in many European countries, providing employment in agriculture as well as in food processing and confectionary production. The production and consumption of black currant products has recently been increasing in Poland, Germany, France, the Baltic States, Great Britain and Switzerland.

In terms of accumulated amounts of ascorbic acid, black currant berries are only surpassed by actinidia and rosehip fruits. In currant berries, the concentration of ascorbic acid varies from 100 to >250 mg/100 g of fruit (Mage, 1993; Brennan, 1996). In certain berries of wild species (*R. nigrum* spp. *Sibiricum*), the amount of ascorbic acid reaches up to 800 mg/100 g (Brennan, 1990). Some researchers state that the ascorbic acid in black currant berries stimulates the antioxidant activity of polyphenolic compounds (Lister et al., 2002). The resistance of different cultivars to abiotic factors and changes in bioactive substances in their berries during ripening-cropping were analysed under the conditions of the Lithuanian climate in different growing years from 2001-2008. Among the twenty-four cultivars analysed, the highest amount of ascorbic acid (vitamin C) was found in the berries of the long-vegetation cultivar 'Vakariai': 267 mg/100 g of berries (Fig. 1). A very high concentration of vitamin C was found to be characteristic of six other black currant cultivars: 'Tiben'<'Joniniai'<'Minaj Shmyriov'<'Ben Moore'<'Ceres'<'Ben Hope'. The amount of ascorbic acid in the above-listed cultivars ranged from 209 to 251 mg/100 g of fruit. The second group in terms of vitamin content included the following cultivars: 'Almiai'<'Ben Dorain'<'Bona'<'Tisel'<'Ben Alder'<'Ben Tirran'<'Ruben'. The ascorbic acid concentration in the berries of these cultivars ranged from 153 to 179 mg/100 g. The majority of the black currant cultivars analysed accumulated from 112 to 140 mg/100 g of vitamin C.

The ascorbic acid content in berries depends on their growing location. Higher levels of ascorbic acid were detected in the berries of cultivars introduced in Northern countries ('Minaj Shmyriov', 'Titania', 'Ojebyn', 'Ben Alder', 'Ben Lomond' and 'Ben Tirran') compared with cultivars from countries in Southern regions (Kampuse et al., 2002; Franceschi et al., 2002; Franke et al., 2004). Depending on their growing location, some black currant cultivars are characterised by higher fluctuations in ascorbic acid levels. For instance, the amount of ascorbic acid in 'Ojebyn' berries varies from 91.3 to 241.0 mg/100 g (Pecho et al., 1993; Rubinskienė, Viškelis, 2002). Some new cultivars (e.g. 'Bona', 'Tisel' and 'Tiben') have been bred from Scandinavian black currant varieties, and the content of ascorbic acid reaches up to 250 mg/100 g in the berries of these new cultivars (Żurawicz et al., 2000). Analysis of the variation of ascorbic acid during berry ripening showed that its content in the berries of individual cultivars depended on the stage of ripeness. The dynamics of ascorbic acid in the berries of cultivars of different earliness depended on the physiological properties of the cultivar and (for certain cultivars) on cultivation conditions. For all cultivars analysed, much higher amounts of ascorbic acid were found in underripe berries. A higher concentration of ascorbic acid was established on day 45 following bush blooming in black currant berries of early and medium-early cultivars, when the berry skin turned pink and the berry mass was increasing rapidly. The highest concentration of ascorbic acid in the berries of late black currant cultivars was established on days 50-55 following the initiation of blooming.

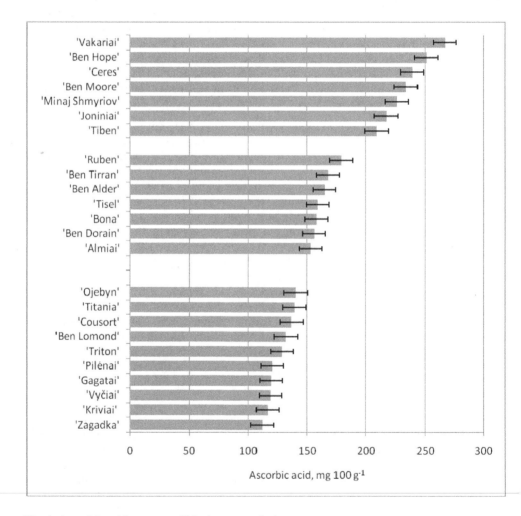

Fig. 1. Ascorbic acid content of black currant fruits

The synthesis of this vitamin is intensive in berries that are not fully ripened, whereas the level of ascorbic oxidase in fruits is low. A sudden decrease in ascorbic acid was recorded in black currants of early and medium-early cultivars that coincided with intense coloration of the berries (dark-brown and black berries). The decrease in the ascorbic acid concentration in berries suitable for consumption can reach 47-52%. The reduction of ascorbic acid during fruit ripening in late black currant cultivars is less pronounced; the ascorbic acid content in the ripe berries of cvs. 'Ben Moore', 'Vakariai', 'Ben Tirran' and 'Ben Alder' was found to be decreased by only 27.3-33.5%. Irrespective of the earliness of the cultivars, overripe berries showed the lowest ascorbic acid contents. The ascorbic acid content in overripe berries was from 14.7 to 29.5% lower. The average correlation (r=-0.52) was found between the ascorbic acid content and the concentration of anthocyanins in black currants. A similar relationship has been noted by other scientists (Brennan, 1990; Trajkovski et al., 2000).

Black currant berries grown under Lithuanian agro-climatic conditions accumulate from 400 to >900 mg of phenolic compounds in 100 g of berries (Fig. 2). The total content of phenolic compounds in most of the analysed cultivars ranged from 600 to 700 mg/100 g. Higher concentrations of phenolic compounds were accumulated in the Scandinavian black currant cultivars 'Ben More', 'Ben Tron', 'Ben Hope', 'Ben Tirran' and 'Tiben'. Additionally, higher amounts of phenolic compounds were found in underripe berries.

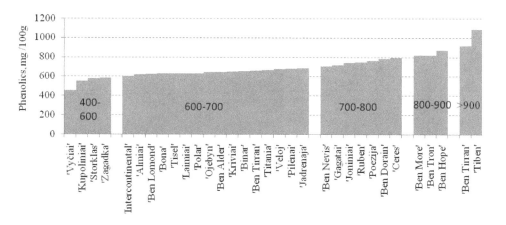

Fig. 2. The total content of phenolic compounds in the berries of different black currant cultivars grown in Lithuania (2005-2010)

The antioxidant activity of ripe berry extracts reaches 78% on average (Fig. 3). The free radical scavenging capacity of ripe and overripe berries was higher than that of underripe berries. A similar tendency was noted by Tabart and co-workers (2006).

Anthocyanins are considered to be powerful (efficient) antioxidants due to their specific structure and activity (Nakajima et al., 2004; Lohachoompol et al., 2004; Galvano, 2005; Fleschhut et al., 2006). Their biological activity has been demonstrated in many studies (Šarkinas et al., 2005; Cooke et al., 2005; Fleschhut et al., 2006).

The persistence of the biological activity of black currant anthocyanin extracts after the recovery procedure was evaluated according to their antibacterial effect (Liobikas et al., 2008). In our opinion, the high antioxidant activity persistence of anthocyanin extracts after recovery procedures indicates the potential of black currant berries to be a good natural source of anthocyanins. In addition, indicators of antioxidant activity and its persistence in recovered anthocyanin extracts support the assumption that these properties depend on the composition of anthocyanin complexes and their aglycones (anthocyanidins) accumulated in the plants of different cultivars (Matsumoto et al., 2002; Blando et al., 2004; Nakajima et al., 2004; Galvano, 2005; Fleschhut et al., 2006).

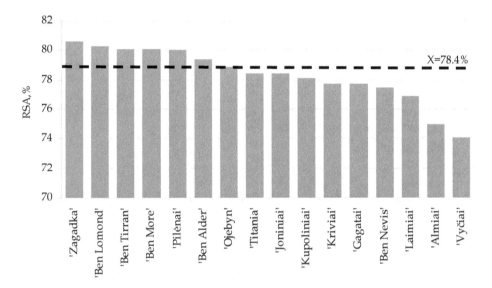

Fig. 3. Radical scavenging capacity of different black currant cultivars

A number of scientists (Kahkonen et al., 2003; Einbond et. al., 2004; Orak, 2006; Tian et al., 2006; Viskelis et al., 2007) are searching for ingredients that can enrich food products with natural antioxidants. Anthocyanins are highly valued as antioxidants and natural colourants (Betsui et al., 2004; Blando et al., 2004; Mazza, 2007). The reported total anthocyanin content of black currants ranges from 214 to 589 mg/100 g, which is 3 to 4 times higher compared to the anthocyanin content found in red raspberries and strawberries and 3 times higher compared with red currants (Mazza, 1993). Black currant berries cultivated under Lithuanian climatic conditions accumulate from 233.5 to 539.7 mg of anthocyanins in 100 g of fruit (Fig. 4). The berries of late and very late black currant cultivars are distinguished by their high amounts of anthocyanins (Rubinskiene et al., 2005). The highest concentrations of these compounds are accumulated in overripe berries. The amount of anthocyanins in overripe berries may be up to 34.5% higher than in ripe berries (Fig. 4). The anthocyanin content in berries depends on cultivar properties and ripening time (Shin et al., 2008; Wang et al., 2009).

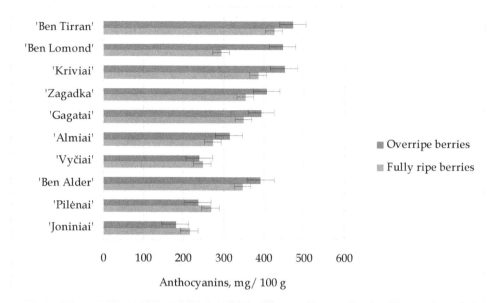

Fig. 4. Anthocyanin content of black currants in ripe and overripe berries

The four main black currant pigments were identified in berries of different maturity; these pigments are cyanidin-3-rutinoside (cy-3-rut), cyanidin-3-glucoside (cy-3glc), delphinidin-3-rutinoside (de-3-rut) and delphinidin-3-glucoside (de-3-glc) (Wrolstad, 2000; Slimestad, Solheim, 2002; Cacace, Mazza, 2003; Määttä-Riihinen et al., 2004). Both the quantitative and qualitative composition of anthocyanin pigments varies during black currant ripening (Rubinskienė et al., 2005; Bordonaba et al., 2010). Cyanidin and delphinidin rutinosides are dominant in ripe berries (Table 1). The amount of cy-3-rut increases considerably during berry ripening. A significantly higher amount of cy-3-rut accumulates in the berries of the 'Almiai' and 'Joniniai' cultivars: 53.08% and 48.33%, respectively. The amounts of anthocyanins in the berry extracts of Lithuanian black currant cultivars are distributed in following sequence: cy-3-rut, 44.14%; de-3-rut, 33.74%; de-3-glc, 11.24%; and cy-3-glc 8.40% (Tab.1). De-3-glc prevails among glucosides in the berries of Lithuanian cultivars. Higher amounts (by 9%) of the dominant pigment cyanidin-3-rutinoside were accumulated in Lithuanian black currant cultivars compared with other black currant berries investigated.

Howard and colleagues indicated that berry weight is closely correlated with the concentration of phenolic compounds in berries (Howard et al., 2003). During the ripening process, a strong negative correlation was detected between berry mass and the amount of anthocyanins accumulated in overripe black currant berries. The mass of overripe berries decreases, while their skin softens, and the concentration of anthocyanins significantly increases. Therefore, all of these quality parameters are correlated with each other (Table 2).

Cultivar	Berry maturity	Anthocyanins, %			
		Cyd-3-rut	Cyd-3-glu	Dpd-3-rut	Dpd-3-glu
'Joniniai'	I	39.06	3.86	48.31	8.77
	II	48.33	4.54	39.11	8.02
'Almiai'	I	36.40	6.06	44.31	13.23
	II	53.08	9.30	28.33	9.29
'Minaj Shmyriov'	I	30.36	4.02	46.49	19.13
	II	43.78	6.45	36.65	13.13
'Vakariai'	I	33.07	4.34	48.68	13.90
	II	38.76	11.30	33.46	16.42
'Ben Alder'	I	30.47	8.62	38.31	22.60
	II	36.63	11.9	31.14	20.33
Mean of cultivars	I	33.87	5.38	45.22	15.51
	II	44.12	8.71	33.74	13.44
$LSD_{05(men\ of\ cultivars)}$		3.002	1.924	1.969	1.641

I-reddish berries, II-black, mature berries

Table 1. Anthocyanin composition (%) in black currant berries of different cultivars;

Rate		Correlation coefficients			
X	Y	Beginning of the ripening	Half ripe	Fully ripe	Overripe
Berry weight, g	Anthocyanins content, mg/100 g	0.18	0.14*	0.46*	-0.63*
Skin strength, N /m2	Anthocyanins content, mg/100 g	-0.15*	0.26	0.54*	0.59*

* - coefficient significant at the probability level of 5%

Table 2. Correlation of the total amount of anthocyanins and physical parameters in black currant berries

4. Raspberry (*Rubus idaeus* L. and *Rubus occidentalis* L.) fruits

Fruits of eight red raspberry cultivars ('Meeker', 'Mirazh', 'Novokitaevskaja', 'Ottawa', 'Sputnica', 'Glen Moy', 'Norna' and 'Siveli'), two yellow raspberry cultivars ('Beglianka' and 'Poranna Rosa') and one black raspberry (*R. occidentalis* L.) cultivar ('Bristol') were investigated in this study.

The pH values of the investigated raspberry fruits varied from 2.96 to 3.35; the content of soluble solids ranged from 10.4 to 12.6 %, and the dry matter content ranged from 12.4 to 19.0 % (Table 3).

Raspberries are a good source of vitamin C (100 g of berries may provide up to 50% of the recommended daily allowance of vitamin C). In this study, the content of ascorbic acid in the investigated raspberry cultivars was found to range from 16.0 (cv. 'Bristol') to 32.1 mg/100 g f.w. (cv. 'Glen Moy') (Tab.3). These data are in agreement with data previously reported for raspberries by other researchers (Haffner et al., 2002; Pantelidis et al., 2007).

Phenolic compounds are the major group of phytochemicals found in berry fruits (Beattie et al., 2005). The total content of phenolics in the investigated raspberries ranged from 223.6 mg/100 g (cv. 'Poranna Rosa') to 690.5 mg/100 g (cv. 'Bristol') (Table 4). The black raspberry *R. occidentalis* exhibited a 1.7-fold higher content of total phenolics compared to the mean value of the investigated *R. idaeus* cultivars (408.8 mg/100 g).

It has previously been reported that the major phenolic compounds found in raspberry fruits are ellagitannins and anthocyanins, whereas hydroxycinnamic acids and flavonols (mainly quercetine glycosides) are the minor phenolic constituents of raspberries (Määttä-Riihinen et al., 2004b).

Cultivar	Dry matter %	Soluble solids %	Ascorbic acid mg/100 g f.w.	pH
Meeker	15.2±0.02e	11.3±0.14b	19.4±0.60bc	3.23±0.05g
Mirazh	15.2±0.11e	12.6±0.06h	19.6±1.90c	2.96±0.02a
Novokitaevskaja	13.2±0.32b	11.0±0.09f	21.4±1.60d	3.35±0.06i
Ottawa	15.7±0.60f	11.5±0.15c	27.2±2.00g	3.07±0.01bc
Sputnica	13.8±0.20c	12.0±0.01g	24.5±1.50ef	3.14±0.00f
Glen Moy	15.4±0.40ef	10.4±0.04a	32.1±0.50h	3.05±0.02b
Norna	12.9±0.04b	11.2±0.14b	25.1±2.00f	3.09±0.04cde
Siveli	14.2±0.22d	10.7±0.11d	18.3±1.80b	3.12±0.01ef
Poranna Rosa	12.4±0.30a	11.2±0.15b	22.0±2.20d	3.10±0.03def
Beglianka	14.1±0.14cd	10.9±0.03e	23.7±2.10e	3.06±0.05bc
Bristol	19.0±0.46g	11.6±0.05c	16.0±0.30a	3.30±0.06h

Values are the mean of three independent determinations ± standard deviation. Different letters in the same raw indicate significant differences (P≤0.05) between cultivars.

Table 3. Content of dry matter, soluble solids, ascorbic acid and pH of raspberry fruits

Cultivar	Free ellagic acid	Ellagitannins	Phenolics	Anthocyanins	RSC, %
			mg/100g f.w.		
Meeker	4.3±0.11c	236.1±12.35f	393.0±14.92d	50.3±2.40e	54.3±0.50c
Mirazh	3.5±0.10e	242.6±13.11c	410.0±15.10f	55.9±2.90b	54.0±2.85c
Novokitaevskaja	5.2±0.14i	219.3±9.96e	366.8±12.67c	56.9±2.00b	53.1±1.35bc
Ottawa	3.4±0.13b	248.9±14.00d	441.2±13.65b	43.6±2.50d	59.3±0.84de
Sputnica	4.7±0.12h	226.2±11.50b	444.8±14.79b	81.7±3.00c	58.6±1.80d
Glen Moy	3.1±0.07d	309.0±16.45g	525.8±19.00i	111.7±4.50f	63.3±2.54f
Norna	4.0±0.12f	225.3±10.27b	420.0±17.10g	82.2±2.20c	58.2±2.61d
Siveli	4.3±0.12c	242.8±14.14c	462.2±10.98h	56.4±3.00b	60.7±0.77e
Poranna Rosa	2.9±0.07a	147.8±10.24a	223.6±9.21a	3.5±0.02a	33.6±1.34a
Beglianka	3.4±0.05b	250.8±9.08d	400.7±16.50e	2.0±0.00a	52.3±0.63b
Bristol	4.1±0.13g	314.1±13.71h	690.5±18.40j	330.8±6.90g	81.5±0.70g

Values are the mean of three independent determinations ± standard deviation. Different letters in the same raw indicate significant differences (P≤0.05) between cultivars.

Table 4. Content of free ellagic acid, ellagitannins, total phenolics, total anthocyanins, and radical scavenging capacity of raspberries

Two major ellagitannins (sanguiin H-6 and lambertianin C) and various ellagic acid derivatives (acylated and/or glycosylated ellagic acid moieties) have been identified in raspberries. However, the free ellagic acid levels are generally low in *Rubus* fruits, and most of the ellagic acid present is in the form of water-soluble ellagitannins within the vacuoles of plant cells (Mullen et al., 2003; Määttä-Riihinen et al., 2004b; Beekwilder et al., 2005). Ellagitanins can be hydrolysed with acids or bases to release hexahydroxydiphenoyl units, which spontaneously form ellagic acid; this reaction is commonly used for the detection and quantification of food ellagitannins (Häkkinen et al., 2000; Vrhovsek et al., 2008; Vekiari et al., 2008; Koponen et al., 2007).

Ellagic acid has received much attention because it has been shown to act as a potent chemopreventive agent (Stoner et al., 2008). Furthermore, ellagic acid/ellagitannins have been found to exhibit antiviral, antibacterial and vasorelaxation properties (Corthout et al., 1991; Goodwin et al., 2009; Nohynek et al., 2006; Mullen et al., 2002).

In this study, the concentration of free (non-tannin) ellagic acid in raspberry fruits ranged from 2.9 to 5.2 mg/100 g in cvs. 'Poranna Rosa' and 'Novokitaevskaja', respectively (Table 4).

The contents of ellagitannins reported in the literature for raspberries are variable, ranging from 20.7 to 329.6 mg/100 g f.w. (De Ancos et al., 2000; Häkkinen et al., 2000; Anttonen and Karjalainen, 2005; Koponen et al., 2007; Vrhovsek et al., 2008). This variation is partly due to the differences in the applied extraction conditions and hydrolysis procedures. However, the reported ellagitannin concentrations for raspberries are generally lower in earlier studies compared to more recent results. The reason for this is that in the earlier studies, ellagic acid was quantified as the only ellagitannin hydrolysis product. However, it was later observed that berry ellagitannins were degraded by acid hydrolysis into ellagic acid and one additional, less polar, derivative (Määttä-Riihinen et al., 2004b). More recently, Vrhovsek and co-workers (2006) identified two other ellagic acid derivatives in addition to ellagic acid after acid hydrolysis of raspberry ellagitannins. Therefore, in later studies, other conversion products of ellagitannin hydrolysis beside ellagic acid were also taken into account.

In the present study, ellagitannins were determined as ellagic acid equivalents after acid hydrolysis. The method previously described by Koponen and co-workers (2007) was applied for ellagitannin analysis. Three major ellagitannin hydrolysis products (ellagic acid, methyl sanguisorboate and methyl gallate) previously identified by Vrhovsek and co-workers (2006) were taken into account. The content of ellagitannins in the investigated raspberries varied from 147.8 to 314.1 mg/100 g f.w. ('Poranna Rosa' and 'Bristol', respectively) (Tab. 4.2). A strong positive correlation (r=0.92) was found between the total content of phenolics and ellagitannin content, showing that raspberry cultivars with a higher phenolic content have higher amounts of ellagitannins.

Red raspberries contain a wide spectrum of anthocyanins, and the major constituents are cyanidin-3-sophoroside, cyanidin-3-glucosylrutinoside and cyanidin-3-glucoside; smaller quantities of cyanidin-3-xylosylrutinoside, cyanidin-3,5-diglucoside, cyanidin-3-rutinoside, cyanidin-3-sambubioside, pelargonidin-3-sophoroside, pelargonidin-3-glucosylrutinoside, pelargonidin-3-glucoside and pelargonidin-3-rutinoside are also present (Mullen et al., 2002; Beekwilder et al., 2005; Kassim et al., 2009; Borges et al., 2010). Significant differences were reported within cultivars with respect to the relative amounts of individual anthocyanins (Beekwider et al., 2005). Moreover, significant year-to-year variations in the contents of anthocyanins in raspberries were noted by different authors (Koponen et al., 2007; Kassim et al., 2009).

Black raspberries accumulate considerably higher amounts of total phenolics, especially anthocyanins, than red raspberries (Wada and Ou, 2002; Weber et al. 2008; Cheplick et al.

2007) blackberries (Wang and Lin, 2000; Wada and Ou, 2002) and some black currant and blueberry genotypes (Moyer et al., 2002). The anthocyanins in black raspberry (*R. occidentalis* L.) consist mainly of cyanidin-3-rutinoside and cyanidin-3-xylosylrutinoside, and smaller amounts of cyaniding-3-glucoside, cyaniding-3-sambubioside and pelargonidin-3-rutinoside have been observed (Tian et al. 2006; Tulio et al. 2008).

In this study, the total anthocyanin content of the investigated red raspberries ranged from 43.6 to 111.7 mg/100 g f.w. (cvs. 'Ottawa' and 'Glen Moy', respectively) (Table 4). Our results are in the range previously reported for raspberries by other scientists (Wada and Ou, 2002; Weber et al. 2008).

Weber and co-workers (2008) analysed 64 raspberry genotypes and noted that some of the red raspberry genotypes presented quite low total anthocyanin contents (up to 40 mg/100 g) in comparison with others (up to 100 mg/100 g). A similar tendency was noted by Lugasi and co-workers (2011); they grouped red raspberry cultivars based on those exhibiting less than 45 mg/100 g of anthocyanins or greater than 55 mg/100 g of anthocyanins. Similarly, it was noted in our study that several of the investigated red raspberry cultivars ('Glen Moy', 'Norna' and 'Sputnica') presented higher total anthocyanin contents than the other red-fruited cultivars (Table 4).

The total anthocyanin content is one of the main criteria used to differentiate between the species *Rubus idaeus* and *Rubus occidentalis* (Suthanthangjai et al., 2005). In the current study, the concentration of anthocyanins in the back raspberry cv. 'Bristol' (330.8 mg/100 g) was almost five times higher than the mean value found in the investigated red-fruited cultivars (67.3 mg/100 g) (Table 4). The concentration of anthocyanins in black raspberries was similar to that reported by other researchers for black raspberries (Wang and Zheng, 2005; Weber et al. 2008).

Due to the presence of recessive genes that suppress the production of anthocyanin pigments, yellow raspberry cultivars contain very low amounts of pigments. In this study, the anthocyanin contents in the yellow raspberry cultivars were found to be 3.5 and 2.0 mg/100 g f.w. (cvs. 'Poranna Rosa' and 'Beglianka', respectively) (Table 4). Similar results for yellow raspberries were reported by other researchers (Anttonen and Karjalainen 2005; Pantelidis et al., 2006; Lugasi et al., 2011).

The amount of anthocyanins represented 47.9% of the amount of total phenolics in black raspberries, whereas in red raspberries, the anthocyanins constituted much lower percentages from 9.9 to 21.2%. This indicates that anthocyanins are not the major phenolic compounds found in red raspberries.

Phytochemicals and vitamins with antioxidant properties are considered to be largely responsible for the health–promoting properties of fruits and vegetables (Stoner et al., 2008). To a significant degree, the antioxidant effects of berry fruits are due to their high content of polyphenols (Szajdek and Borowska, 2008). The contribution of vitamin C to the antioxidant activity of raspberries seems to be relatively low; according to Beekwiler and co-workers (2005), the contribution of vitamin C to the total antioxidant activity of red raspberries was approximately 20%, and Borges and co-workers (2010) reported that the contribution of vitamin C to the detected antioxidant capacity of raspberries was 11%.

The radical scavenging capacity (RSC) of the investigated raspberries in the DPPH[•] reaction system ranged from 33.6 to 81.5% (cvs. 'Poranna Rosa' and 'Bristol', respectively) (Table 4). The RSC of black raspberry cv. 'Bristol' was 30% higher than the mean RSC value of investigated *R. idaeus* cultivars.

The RSC of the investigated raspberry cultivars was highly correlated with their total phenolics content (r= 0.99). This is in agreement with previously reported results showing that the antioxidant properties of berries correlate well with their phenolic content (Moyer et al., 2002; Anttonen and Karjalainen, 2005; Pantelidis et al., 2007; Poiana et al., 2010). Additionally, the RSC of raspberry fruits was highly correlated with their ellagitannin content (r=0.89), implying that the antioxidant activity of raspberries is largely due to the presence of ellagitannins. These findings are in agreement with previous reports (Mullen et al., 2002; Beekwider et al., 2005; Borges et al., 2010). In this study, a medium correlation (r=0.60) was found between the total anthocyanin content of the eight investigated red-fruited raspberries and their RSC. Beekwilder and co-workers (2005) reported that anthocyanins are responsible for approximately 25% of the antioxidant capacity of red raspberry fruits. Borges et al. (2010) found that red raspberry anthocyanins were responsible only for 16% of the total antioxidant capacity. This indicates that anthocyanins are not the dominant antioxidants in red raspberries. In contrast, for black raspberries, anthocyanins have been reported to be the major phenolic antioxidants (Tulio, 2008).

The fruit colour of raspberries is determined not only by the combination of the contents of anthocyanins present but also by the cellular environment in which the anthocyanins are suspended. Jennings and Carmichael (1980) have investigated the inheritance of anthocyanins in various *Rubus* species; these authors noted that when pelargonidin pigments were incorporated, some degree of orange colour was imparted to progenies. Melo et al. (2002) have shown that the colouration of raspberries is not based on co-pigmentation but is due mainly to effects of the pH in vacuoles. Because of these pH effects, berries with the same anthocyanin constitution may exhibit very different fruit colours.

Cultivar	L*	a*	b*	C	h°
Meeker	37.29±0.17d	28.11±0.40d	9.57±0.23e	29.64±0.36d	18.80±0.55d
Mirazh	34.95±0.28b	25.95±0.13c	8.84±0.14d	27.41±0.10c	18.81±0.35d
Novokitaevskaja	34.81±0.09b	23.64±0.20b	7.48±0.12c	24.79±0.21b	17.55±0.22b
Sputnica	32.26±0.23c	20.26±0.38a	5.99±0.29a	21.13±0.42a	16.46±0.57a
Glen Moy	31.68±0.19a	20.40±0.47a	6.50±0.44b	21.41±0.58a	17.64±0.75b

Parameters: L* (lightness), a* (from (-) green to (+) red), b* (from (-) blue to (+) yellow), h°=arctan (b*/a*), C= [(a*)2+(b*)2] 0.5.
Values are the mean of three independent determinations ± standard deviation. Different letters in the same raw indicate significant differences (P≤0.05) between cultivars.

Table 5. Colour determination values (L*, a*, b*, C, h°) of raspberry fruits

The colour of five raspberry cultivars was measured by the CIEL*a*b* method. The estimated CIEL*a*b* values are given in Table 5. A very strong negative correlation (r =-0.89) was found between the total anthocyanin content of the raspberries and the lightness coordinate L*. Slightly lower negative correlations were found between the total anthocyanin content and the redness coordinate a* as well as between the total anthocyanins content and the chroma (colour saturation) parameter C (r=-0.84 and -0.83, respectively). The correlations between the total anthocyanin content and the yellowness coordinate b* and the hue angle h° of raspberries was -0.77 and -0.53, respectively. According to our results, the L* colour parameter was sufficiently accurate for screening the anthocyanin concentration of raspberries. It should be noted that the colour of homogenised berries was measured in this study. It has previously been shown that when the colour of the surface of the raspberry fruit was measured, the colour parameters were not good indicators of the anthocyanin levels of raspberries (Haffner et al., 2002).

5. Black chokeberry (*Aronia melanocarpa*) fruits

Black chokeberries accumulate extremely high amounts of anthocyanins and other polyphenolic substances (Valcheva-Kuzmanova et al. 2004; Oszmiański, Wojdylo 2005). *Aronia melanocarpa* represents one of the richest natural sources of anthocyanins, containing from 300 to 630 mg of anthocyanins in 100 g of berries. Black chokeberries were reported to accumulate higher amounts of anthocyanins as compared to black currants, blackberries and elderberries (Oszmianski, Sapis, 1988; Benvenuti et al., 2004). A strong correlation was found between fruit antioxidant activity and the total amount of polyphenolic substances (Bermúdez-Soto, Cevallos-Casalsem et al., 2006). The results of various studies support the positive impact of phenolic antioxidants on human health (Borowska et al., 2005). Currently, most European juice manufacturers include black chokeberry juice among their products, and the demand for black chokeberry concentrate is increasing. According to earlier investigations, the anthocyanins present in *Aronia melanocarpa* are a mixture of four different cyanidin glucosides: 3-galactoside, 3-glucoside, 3-arabinoside and 3-xyloside, of which cyanidin 3-galactoside is the major component (Oszmianski, Sapis, 1988; Wu et al., 2004; Jakobek et al., 2007). In our study, the fruits of black chokeberry cultivars accumulated from 634.6 ('Aron') to 868.9 mg/100 g ('Viking') of anthocyanins (Table 6). The amount of anthocyanins in overripe fruits of *Aronia melanocarpa* was lower by 5.3 to 38.5% (in cvs. 'Aron' and 'Viking', respectively). The highest amount of phenolic compounds, 3647.0 mg/100 g, was found to accumulate in fruits of cv. 'Viking'. As was observed by Rop and colleagues (2010) while investigating five cultivars of *Aronia melanocarpa* of different origins, 'Viking' berries are distinguished with respect to the abundance of the mentioned compounds. However, the amount of total phenolics in overripe berries of cv. 'Viking' was lower by 28% compared to ripe fruits (Table 6). This tendency was also observed by Parr and colleagues (2000), who analysed the variations of phenolic compounds in overripe garden plants. It is estimated that the total phenolic concentration in fruits decreases due to increased polyphenolic oxidase activity. Different authors reported detecting high amounts of total phenolics in black chokeberries, and some reported concentrations that were even higher than those found in our study (Benvenuti et al., 2004; Oszmianski and Wojdylo, 2005). Based on the examination of the phenolic contents and antioxidant activity of 92 plant extracts, Kahkonen and co-workers (1999) reported that among the edible plant materials investigated, remarkably high antioxidant activities and total phenolic contents were found in berry fruits, especially in black chokeberries (Walther, Schnell, 2009). The radical scavenging capacity of the analysed *Aronia melanocarpa* berries ranged from 83.6-86.7%, though the extracts of overripe berries scavenged only slightly greater amounts of free radicals (Table 6).

Cultivar	Berry maturity	Total anthocyanins content, mg/100 g	Total phenolic content, mg/100 g	RSC, %
'Viking'	I*	868.9	3647.0	84.1
	II**	534.4	2609.5	86.7
'Aron'	I	634.6	3123.0	84.1
	II	601.2	3028.7	84.8
Aronia var.	I	701.4	3028.7	83.6
cleata	II	617.9	3028.7	84.2
LSD$_{05}$		28.49	153.88	3.23

*fullyripe fruit
** overripe fruit

Table 6. Chemical composition and radical scavenging capacity of black chokeberries

The colour parameters of berries and homogenised berry purees are shown in Table 7. 'Viking' berries and their homogenised purees were distinguished based on their colour intensity. Negative values of the colour coordinate b* support the conclusion that greater amounts of anthocyanin pigments are accumulated in chokeberry skins. The pulp tissues present in berry puree changed a value of coordinate b* into positive side. Substantial differences between the colour coordinates a* and b* in berries and purees have a marked influence the hue angle (h°) values (Table 7). The variations of the hue values are the most prominent compared to those of the other analysed colour quality parameters. The h° values of homogenised berry purees did not show a significant difference.

Cultivar	Preparation	L*, %	a*	b*	C	h°
'Viking'	fruit	24.86	0.31	-1.02	1.1	286.9
	puree	26.40	3.95	0.96	4.1	13.7
'Aron'	fruit	26.67	0.67	-0.24	0.7	340.3
	puree	26.53	4.38	1.03	4.5	13.2
Aronia	fruit	21.59	0.53	-0.04	0.5	355.7
var. cleata	puree	26.52	4.20	1.09	4.3	14.5
LSD$_{05(fruit)}$		1.22	0.03	0.02	0.04	16.38
LSD$_{05(puree)}$		1.32	0.21	0.05	0.22	0.69

Table 7. Colour determination values of black chokeberries and homogenised purees

6. Elderberry (*Sambucus nigra* L.) fruits

Elderberries are one of the richest sources of anthocyanins, and the content of anthocyanins in elderberries has been reported to be as high as 1000 mg/100 g (Bronnum-Hansen, Flink, 1986; Kaack, Austed, 1998). Direct evidence has only recently been provided showing that anthocyanins can be absorbed by humans. Cao and Prior (1999) showed that after oral administration of elderberry extract, cyanidins are absorbed in their glycosidic forms. Additional evidence is available concerning the absorption of anthocyanins, including cyanidin 3-glucoside, by mammals (Talavéra et al., 2004; Galvano et al., 2007). In this study, the investigated elderberry cultivars (overripe berries) were found to present high amounts of anthocyanins (709.7 mg/100 g, on average) (Table 8). However, elderberries accumulated 3% less anthocyanins on average compared to *Aronia melanocarpa* fruits (see Tables 8 and 6). Higher amounts of total anthocyanins and total phenolics were found in *Sambucus nigra* cv. 'Lacimiata' (Table 8).

The RSC of *Sambucus nigra* fruits was lower when compared with the RSC of black raspberry and *Aronia melanocarpa* cultivars (Viškelis et al., 2010). Kaack and Aused (1998) reported large differences (from 6 to 60 mg/100 g) in the ascorbic acid content in elderberries depending on the conditions in which the berries were grown. In our study, the average content of ascorbic acid in fruits of *Sambucus nigra* was 29.5 mg/100 g (Table 8).

Cultivar	Ascorbic acid, mg/100g	Anthocyanins, mg/100g	Phenolic compounds, mg/100g	RSC, %
Sambucus nigra 'Aurea'	30.8	505.0	800.2	66.6
Sambucus nigra 'Lacimiata'	28.2	914.3	1050.0	62.9

Table 8. Chemical composition and radical scavenging capacity of elderberry fruits

7. Sea buckthorn (*Hippophae rhamnoides* L.) fruits

Sea buckthorn is becoming increasingly popular due to its valuable fruits, which are suitable for nutritional and medical purposes (Li, Schroeder, 1996). Sea buckthorn fruits are an excellent source of bioactive phytochemicals, such as carotenoids, tocopherols, sterols, vitamin C, organic acids, and polyphenols (Arimboor et al., 2006; Ranjith et al., 2006). Food, medicine, veterinary and cosmetic industries commonly uses sea buckthorn in their products (Li, 2002; Li, Zeb, 2004). Sea buckthorn varies widely both between populations and between individuals within the same population (Yao, Tigerstedt, 1994; Tang, Tigersted, 2001). The content of valuable fruit constituents within populations also varies (Beveridge et al. 1999; Yang, Kallio et al. 2002). From a practical point of view, cultivated sea buckthorn varieties are more valuable because plants of selected genotypes are adapted to certain environmental conditions; their fruits contain higher amounts of biologically active substances. Mechanical harvesting remains a main problem in developing new cultivars (Trajkovski, Jeppson, 1999).

Valuable sea buckthorn cultivars were selected at the Altai Horticulture Institute (Kalinina, Panteleyeva, 1987). Some of these cultivars were tested at the Lithuanian Institute of Horticulture from 1984-1989. The selected cultivars yielded large fruits (average fruit weight was 0.83 g) that were relatively easily picked by hand. However, numerous plants of Siberian-bred sea buckthorn cultivars have died when grown in orchards. It appears that Lithuanian winters, with thaws and frosts, are detrimental to continental climate-adapted cultivars of sea buckthorn.

Under Lithuanian climate conditions better approved at the Moscow State University Botanical Garden bred sea buckthorn cultivars 'Avgustinka', 'Botanicheskaya', 'Podarok sadu' and 'Trofimovskaya'. The fruits of the tested sea buckthorn cultivars gained colour in the middle of August. The average fruit weight at that point was 0.57 g. Despite showing external signs of maturity, the fruit weight continued to increase and reached a maximum (0.81 g) in the middle of September. A similar tendency of increasing berry weight was noted by Raffo et al. (2004).

The firmness of sea buckthorn fruits decreased as they matured: on the 13th of August, the peel strength was 48.7 N/cm^2 on average, and on the 24th of September, it was almost two times lower (25.8 N/cm^2). Hand picking of sea buckthorn fruit gradually becomes more difficult as the required fruit detachment force usually remains the same, while the skin firmness decreases. Harvesting of sea buckthorn is influenced by weather conditions and should usually be completed in September.

In this study, the highest content of dry matter was found in fruits of cv. 'Trofimovskaya', which presented a 17.2% dry matter content on average (Table 9). The rest of the tested cultivars exhibited significantly lower amounts of dry matter. Changes in the dry matter content during fruit ripening were insignificant.

Cultivar	Dry matter	Soluble solids	Total sugars	Titratable acidity	Vitamin C	Carotenoids
	%				mg/100 g	
Avgustinka	15.7	9.2	3.5	1.9	66.3	15.6
Botanicheskaya	16.0	8.3	2.9	1.6	74.2	10.0
Podarok sadu	15.2	8.9	4.1	1.9	74.9	15.2
Trofimovskaya	17.2	9.7	4.2	1.7	109.3	15.2
LSD$_{05}$	0.91	0.51	0.80	0.17	13.09	3.84

Table 9. Biochemical composition of sea buckthorn fruits

The content of soluble solids varied from 2.85 to 22.74% depending on the origin and stage of maturity of the plants and the climate under which they were cultivated (Raffo et al., 2004; Antonelli et al., 2005; Dwivedi et al., 2005; Tiitinen et al., 2005). In this study, highest contents of soluble solids were found in fruits of cvs. 'Trofimovskaya' and 'Avgustinka' (9.7 and 9.2%, respectively) (Table 9). Lower precipitation and higher temperatures result in higher dry matter and soluble solid contents.

The highest content of total sugars was found in fruits of cvs. 'Podarok sadu' and 'Trofimovskaya' (4.1 and 4.2%, respectively) (Table 9). The total sugar content in ripening fruits increased initially and later remained stable or slightly decreased.

Sea buckthorn fruits are among the richest food sources of vitamin C (Beveridge et al. 2002). Under Lithuanian agroclimatic conditions, the greatest amount of vitamin C accumulated in fruits of cv. 'Trofimovskaya', with a content of 109.3 mg/100 g (Table 9). High vitamin C content (82 mg/100 g) in fruits of 'Trofimovskaya' was also reported by Univer et al. (2004). The vitamin C concentration ranges from 28 to 310 mg in 100 g of berries in the European subspecies *rhamnoides* (Rousi, Aulin, 1977; Jeppsson, Gao, 2000; Yao et al., 1992). Wild fruits of subsp. *Sinensis* (native to China) contain 5 - 10 times greater amounts of vitamin C than the fruits of the European subsp. *rhamnoides* (Kallio et al., 2002). In our study, the content of vitamin C decreased during fruit ripening. The same tendency was reported by Univer et al. (2004).

The highest titratable acidity was found for cvs. 'Avgustinka' and 'Podarok sadu' and was 1.9 % for both cultivars, whereas the lowest was observed for cv. 'Botanicheskaya' (1.6 %) (Table 9).

The lowest content of carotenoids (10.0 mg/100 g) was found in fruits of cv. 'Botanicheskaya' (Table 9). The other tested cultivars accumulated up to 15.6 mg/100 g of carotenoids. Andersson et al. (2009) established that the carotenoid content increased during sea buckthorn fruit ripening, exhibiting values of 1.5-18.5 mg/100 g of fresh weight depending on the cultivar, harvest time, and year.

The highest content of total phenolics of 237.9 mg/100 g was found in fruits of cv. 'Trofimovskaya', whereas fruits of cv. 'Avgustinka' exhibited the lowest content of total phenolics of 179.7 mg/100 g (Fig. 5). The radical scavenging capacity of cv. 'Trofimovskaya' was the highest among the investigated sea buckthorn cultivars (27.4%) (Fig. 6). Sea buckthorn possesses antioxidant properties due to its ascorbic acid, carotenoid and polyphenolic compound content (Ecclesten et al., 2002; Zadernowski et al. 2003; Cenkowski et al., 2006). Sea buckthorn generally shows lower antioxidant activity in DPPH• reaction system compared with some other berries and small fruits (Li et al., 2009; Viskelis et al., 2010). It should be noted that the antioxidant activity of the lipophilic berry fraction was not evaluated in our study. Gao and co-workers (2000) reported that lipophilic sea buckthorn berry fractions were most effective if the comparison was based on the ratio between the antioxidant capacity and content of antioxidants. It has been reported that sea buckthorn berry juice has a higher lipophilic antioxidant capacity than tomato, carrot or orange juice (Müller et al., 2011).

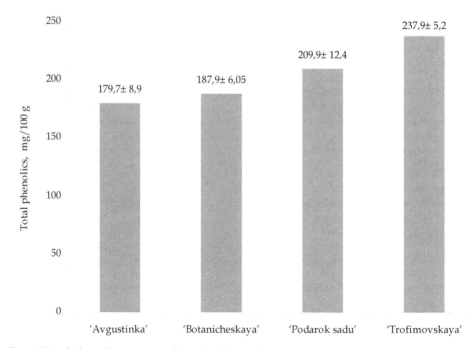

Fig. 5. Total phenolics content of sea buckthorn fruits

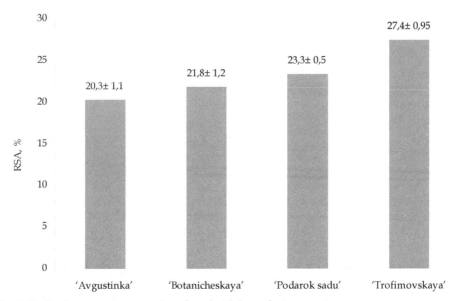

Fig. 6. Radical scavenging capacity of sea buckthorn fruit extract.

8. Conclusions

The results obtained in this study show that the chemical composition and antioxidant capacity of investigated berry fruits varies significantly due to genetic factors. Based on phenolic compounds concentration the investigated berry species may be ranked as follows: *Aronia melanocarpa>Sambucus nigra* L.*>Ribes nigrum* L.*>Rubus occidentalis* L.*>Rubus idaeus* L.*> Hippophae rhamnoides* L. Antioxidant capacities varied significantly among berries and cultivars investigated in this study and were highly correlated with phenolic compounds content. Among all berry species tested, fruits of *Aronia melanocarpa* were found to have the highest amounts of anthocyanins, followed by *Sambucus nigra* L., *Ribes nigrum* L., and black raspberry *Rubus occidentalis* L. The highest ascorbic acid content among investigated berry species was found in black currant fruits, followed by the sea buckthorn fruits.

The results indicate that analyzed berry species are rich sources of biologically active substances and posses potent antioxidant activities. The study expands the knowledge about variation in the content of biologically active compounds in different berry species.

9. References

Andersson, S.C.; Olsson, M.E., Johansson, E., Rumpunen, K. (2009). Carotenoids in sea buckthorn (*Hippophae rhamnoides* L.) berries during ripening and use of pheophytin a as a maturity marker. *Journal of Agricultural and Food Chemistry*, Vol.57, No.1, pp. 250-258, ISSN 0021-8561.

Antonelli, M.; Raffo, A., Paoletti, F. (2005). Biochemical changes during ripening of seabuckthorn (*Hippophae rhamnoides*) fruits, In: *Seabuckthorn (Hippophae L.) A Multipurpose Wonder Plant*, V. Singh, (Ed.), Vol.2, 285–312, Daya Publishing House, New Delhi, India.

Anttonen, M. J.; Karjalainen, R.O. (2005). Environmental and genetic variation of phenolic compounds in red raspberry. *Journal of Food Composition and Analysis*, Vol.8, pp. 759-769, ISSN 0889-1575.

AOAC (1990). Vitamin C (ascorbic acid) in vitamin preparations and juices, In: *Official Methods of Analysis,15thed.*, Helrich K., (Ed.), 1058, AOAC Inc., Arlington, VA.

Arimboor, R.; Venugopalan, V.V., Sarinkumar, K., Arumughan, C., Sawhney, R.C. (2006). Integrated processing of fresh Indian sea buckthorn (*Hippophae rhamnoides*) berries and chemical evaluation of products. *Journal of the Science of Food and Agriculture*, Vol.86, pp. 2345-2353, ISSN 1097-0010.

Artemio, Z.; Tulio, J.R., Reese, R. N., Whyzgoski, F. J., Rinaldi, P. L., Fu, R., Scheerens, J.C., Miller, A.R. (2008). Cyanidin 3-rutinoside and cyaniding 3-xylosylrutinoside as primary phenolic antioxidants in black raspberry. *Journal of Agricultural and Food Chemistry*, Vol.56, pp. 1880-1888, ISSN 0021-8561.

Beattie, J.; Crozier, A., Duthie ,G.G. (2005). Potential health benefits of berries. *Current Nutrition & Food Science*, Vol.1, pp. 71-86, ISSN 1573-4013.

Beekwilder, J.; Jonker, H., Meesters, P., Hall, R.D., van der Meer, I.M., Rick de Vos, C.H. (2005). Antioxidants in raspberry: on-line analysis links antioxidant activity to a diversity of individual metabolites. *Journal of Agricultural and Food Chemistry*, Vol.53, pp. 3313-3320, ISSN 0021-8561.

Benvenuti, S.; Pellati, F, Melegari, M, Bertelli, D. (2004). Polyphenols, anthocyanins, ascorbic acid, and radical scavenging activity of Rubus, Ribes and Aronia. *Journal of Food Science*, Vol.69, pp. 164-169, ISSN 1750-3841.

Betsui, F.; Tanaka–Nishikawa, N., Shimomura, K. (2004). Anthocyanin production in adventitious root cultures of *Raphanus sativus* L. cv. Peking Kouskin. *Plant Biotechnology*, Vol.21, pp. 387–391, ISSN 1467-7652.

Beveridge, T.; Harrison, J.E., Drover, J. (2002). Processing effects on the composition of sea buckthorn juice from *Hippophae rhamnoides* L. cv. Indian summer. *Journal of Agricultural and Food Chemistry*, Vol.50, pp. 113-116, ISSN 0021-8561.

Beveridge, T., L,i T.S.C., Oomah, B.D., Smith, A. (1999). Sea buckthorn products: manufacture and composition. *Journal of Agricultural and Food Chemistry*, Vol.47, pp. 3480-3488, ISSN 0021-8561.

Blando, F.; Gerardi, C. Nicoletti, I. (2004). Sour Cherry (*Prunus cerasus* L.) anthocyanins as ingredients of functional food. *Journal of Biomedicine and Biotechnology*, Vol.5, pp. 253-258, ISSN 11107243.

Bordonaba, J.G.; Gemma, A., Chope, G.A. and Terry, L.A. (2010). Maximising blackcurrant anthocyanins: Temporal changes during ripening and storage in different genotypes. *Journal of Berry Ressearch*, Vol.1, pp. 73-80, ISSN 1878-5093.

Borges, G.; Degeneve, A., Mullen, W., Crozier, A. (2010). Identification of flavonoid and phenolic antioxidants in black currants, blueberries, raspberries, red currants and cranberries. *Journal of Agricultural and Food Chemistry*, Vol.58, pp. 3901-3909, ISSN 0021-8561.

Borowska, E.J.; Szajdek, A., Borowski, J., (2005). Antioxidant properties of fruits, vegetables and their products. *Fruit Processing*, Jan./Feb., pp. 38-43.

Brand-Williams, W.; Cuvelier, M.E., Berset, C. (1995). Use of a free radical method to evaluate antioxidant activity. *LWT-Food Science and Technology*, Vol.28, pp. 25-30, ISSN 0023-6438.

Bravo, L. (1998). Polyphenols: chemistry, dietary sources, metabolism, and nutritional significance. *Nutrition Reviews*, Vol.56, pp. 317-333, ISSN 1753-4887.

Brennan, R. (1996). Currants and gooseberries, In: *Fruit breeding*, J. Janick, & J. N. Moore, (Eds.), Vol.2. Small fruit and vine crops, 191-295, Wiley and Sons, New York.

Brennan, R.M. (1991). Currants and gooseberries (*Ribes*). *Acta Horticulturae*, Vol.290, pp.457-488, ISSN 0567-7572.

Brennan, R.M., Lanham, P.G., McNicol, R.J. (1993). *Ribes* breeding and research in the UK. *Acta Horticulturae*, Vol.352, pp. 267-275, ISSN 0567-7572.

Bronnum-Hansen, K.; Flink, J.M. (1986). Anthocyanin colorants from elderberry (Sambucus nigra L.) IV. Further studies on production of the liquid extracts, concentrates and freeze dried powder. *Journal of Food Technology*, Vol.21, pp. 605-614, ISSN 1993-6036.

Cacace, J. E.; Mazza, G. (2003). Optimization of extraction of anthocyanins from black currants with aqueous ethanol. *Journal of Food Science*, Vol.68, pp. 240-248, ISSN 1750-3841.

Cao, G.; Prior, R.L. (1999). Anthocyanins are detected in human plasma after oral administration of an elderberry extract. *Clinical Chemistry*, Vol.45, pp. 574-576.

Cenkowski, S.; Yakimishen, R., Przybylski, R., Muir, W.E. (2006). Quality of extracted sea buckthorn seed and pulp oil. *Canadian Biosystems Engineering*, Vol.48, pp. 3.9-3.16, ISSN 1492-9058.

Cevallos-Casals, B.; Byrne, D., Okie, W., Cisneros-Zevallos, L. (2006). Selecting new peach and plum genotypes rich in phenolic compounds and enhanced functional properties. *Food Chemistry*, Vol.96, pp. 273-280, ISSN 0308-8146.

Cheplick, S.; Kwon, Y-I., Bhowmik, P., Shetty, K. (2007). Clonal variation in raspberry fruit phenolics and relevance for diabetes and hypertension management. *Journal of Food Biochemistry*, Vol.31, pp. 656-679, ISSN 1745-4514.

Connor, A.M.; Luby, J.J., Tong, C.B.S., Finn, C.E., Hancock, J.F. (2002). Variation and heritability estimates for antioxidant activity, total phenolic content, and anthocyanin content in blueberry progenies. *Journal of the American Society for Horticultural Science*, Vol.127, pp. 82–88, ISSN 0003-1062.

Cooke, D.; Steward, W.P., Gescher, A.J., Marczylo, T. (2005). Anthocyanins from fruits and vegetables – does bright colour signal cancer chemopreventive activity? *European Journal of Cancer*, Vol.41, pp. 1931-1940, ISSN 0959-8049.

Corthout, J.; Pieters, L.A., Claeys, M., Vandenberghe, D.A., Vlietinck, A.J. (1991). Antiviral ellagitannins from *Spondias mombin*. *Phytochemistry*, Vol.30, pp. 1129-1130, ISSN 0031-9422.

De Ancos, B.; González, E.M., Cano, M.P. (2000). Ellagic acid, vitamin C, and total phenolic contents and radical scavenging capacity affected by freezing and frozen storage in raspberry fruit. *Journal of Agricultural and Food Chemistry*, Vol.48, pp. 4565-4570, ISSN 0021-8561.

Dwivedi, S. K.; Singh, R., Ahmed, Z. (2005). Morpho-biochemical characteristics of seabuckthorn (*Hippophae rhamnoides*) growing in cold arid Ladakh Himalayas, In: *Seabuckthorn (Hippophae L.). A Multipurpose Wonder Plant*, V. Singh, (Ed.), Vol.2, 151–158, Daya Publishing House, New Delhi, India.

Ecclesten, C.; Baoru, Y., Tahvonen, R., Kallio, H., Rimbach, G.H., Minihane, A.M. (2002). Effects of antioxidantrich juice (sea buckthorn) on risk factors for coronary heart desease in humans. *The Journal of Nutritional Biochemistry*, Vol.13, pp. 346-354, ISSN 0955-2863.

Einbond, L.; Reynertson, K., Luo, X. et al. (2004). Anthocyanin antioxidants from edible fruits. *Food Chemistry*, Vol.84, pp. 23-28, ISSN 0308-8146.

Fleschhut, J.; Kratzer, F., Rechkemmer, G., Kulling, S.E. (2006). Stability and biotransformation of various dietary anthocyanidins *in vitro*. *European Journal of Nutrition*, Vol.45, pp. 7-18, ISSN 1436-6207.

Franceschi, V. R.; Tarlyn, N. L. (2002). L-Ascorbic acid is accumulated in source leaf phloem and transported to sink tissues in plants. *Plant Physiology*, Vol.130, pp. 649-656, ISSN 0032-0889.

Franke, A. A.; Custer, L. J., Arakak,i C., Murphy, S. P. (2004). Vitamin C and flavonoid levels of fruits and vegetables consumed in Hawaii. *Journal of Food Composition and Analysis*, Vol.17, pp. 1-35, ISSN 0889-1575.

Galvano, F. (2005). The chemistry of anthocyanins. *Functional ingredients magazine*, Vol.7, pp. 1-6.

Galvano, F.; Fauc,i L. L, Vitaglione, P. Fogliano, V., Vanella, L., Felgine,s C. (2007). Bioavailability, antioxidant and biological properties of the natural free-radical scavengers cyanidin and related glycosides. *Annali dell'Istituto superiore di sanita*, Vol.43, pp. 382-393, ISSN 0021-2571.

Gao, X.; Ohlander, M., Jeppsson, N., Bjork, L., Trajkovski, V. (2000). Changes in antioxidant effects and their relationship to phytonutrients in fruits of sea buckthorn (*Hippophae rhamnoides* L.) during maturation. *Journal of Agricultural and Food Chemistry*, Vol.48, pp. 1485-1490, ISSN 0021-8561.

Giusti, M.M.; Wrolstad, R.E. (2001). Characterization and measurement of anthocyanins by UV–visible spectroscopy, In: *Current protocols in food analytical chemistry*, R. E. Wrolstad, T. E. Acree, H. An, E. A. Decker, M. H. Penner, D. S. Reid, S. J. Schwartz, C. F. Shoemaker, & P. Sporns, (Eds.), F1.2.1–F 1.2.13, Wiley, New York, ISBN 9780471142911.

Goodwin, E.C.; Arwood, W.J., DiMaio, D. (2009). High-throughput cell-based screen for chemicals that inhibit infection by simian virus 40 and human polyomaviruses. *Journal of Virology*, Vol.83, pp. 5630-5639, ISSN 1098-5514.

Haffner, K.; Rosenfeld, H.J., Skrede, G., Wang, L. (2002). Quality of red raspberry *Rubus idaeus* L. cultivars after storage in controlled and normal atmospheres. *Postharvest Biology and Technology*, Vol.24, pp. 279-289, ISSN 0925-5214.

Häkkinen, S.H.; Kärenlampi, S.O., Mykkänen, H.M., Heinonen, I.M., Törrönen, A.R. (2000). Ellagic acid content in berries: Influence of domestic processing and storage. *European Food Research and Technology*, Vol.212, pp. 75-80, ISSN 1438-2385.

Howard, L. R.; Clark, J. R. and Brownmiller, C. (2003). Antioxidant capacity and phenolic content in blueberries as affected by genotype and growing season. *Journal of the Science of Food and Agriculture*, Vol.83, pp. 1238–1247, ISSN 1097-0010.

Hummer, K.; Barney, D. (2002). Crop reports. Currants. *HortTechnology*, Vol.12, pp. 377-387, ISSN 1063-0198.

Jakobek, L.; Šeruga, M., Medovidovic-Kosanovic, M., Novak, I. (2007). Antioxidant activity and polyphenols of aronija in comparison to other berry spesies. *Agriculturae Conspectus Scientificus*, Vol.72, No.4, pp. 301-306, ISSN 1331-7776.

Jennings, D.L.; Carmichae,l E. (1980). Anthocyanin variation in the genus *Rubus*. *New Phytologist*, Vol.84, pp. 505-513, ISSN 1469-8137.

Jeppsson, N.; Gao, X.Q. (2000). Changes in the contents of kaempherol, quercetin and L-ascorbic acid in sea buckthorn berries during maturation. *Agricultural and Food Science in Finland*, Vol.9, No.1, pp. 17-22, ISSN 1239-0992.

Joseph, J.A.; Denisova, N.A., Bielinski, D., Fisher, D.R., Shukitt-Hale, B. (2000). Oxidative stress protection and vulnerability in aging: putative nutritional implications for intervention. *Mechanisms of Ageing and Development*, Vol.116, pp. 141-153, ISSN 0047-6374.

Kaack, D.; Austed, T. (1998). Interaction of vitamin C and flavonoids in elderberry (Sambucus nigra L.) during juice processing. *Plant Foods for Human Nutrition*, Vol.52, pp. 187-198, ISSN 1573-9104.

Kahkonen, M.P.; Heinamaki, J., Ollilainen, V., Heinonen, M. (2003). Berry anthocyanins: isolation, identification, and antioxidant activities. *Journal of the Science of Food and Agriculture*, Vol.83, pp. 1403-1411, ISSN 1097-0010.

Kahkonen, M.P.; Hopia, A.I., Vuorela, H.J., Rauha, J.P., Pihlaja, K., Kujala, T.S., Heinonen, M. (1999). Antioxidant activity of plant extracts containing phenolic compounds. *Journal of Agricultural and Food Chemistry*, Vol.47, pp. 3954-3962, ISSN 0021-8561.

Kalinina, I.P.; Panteleyeva, Y.I. (1987). Breeding of sea buckthorn in the Altai, In: *Advantages in Agricultural Science*, Hryukov A.B, (Ed.), 46-87, Moscow, Russia.

Kallio, H.; Yang, B., Peippo, P. (2002). Effects of different origins and harvesting time on vitamin C, tocopherols, and tocotrienols in sea buckthorn (*Hippophaë rhamnoides*) berries. *Journal of Agricultural and Food Chemistry*, Vol.50, pp. 6136-6142, ISSN 0021-8561.

Kallio, H.; Yang, B., Peippo, P., Tahvonen, R., Pan, R. (2002). Triacylglycerols, glycerophospholipids, tocopherols, and tocotrienols in berries and seeds of two subspecies (ssp. *sinensis* and *mongolica*) of sea buckthorn (*Hippophaë rhamnoides*). *Journal of Agricultural and Food Chemistry*, Vol.50, pp. 3004–3009, ISSN 0021-8561.

Kampuse, S.; Kampuss, K., Pizika, L. (2002). Stability of anthocyanins and ascorbic acid in raspberry and blackcurrant cultivars during frozen storage. *Acta Horticulturae*, Vol.585, pp. 507-600, ISSN 0567-7572.

Kassim, A.; Poette, J., Paterson, A., Zait, D., McCallum, S., Woodhead, M., Smith, K., Hackett, C., Graham, J. (2009). Environmental and seasonal influences on red raspberry anthocyanin antioxidant contents and identification of quantitative traits loci (QTL). *Molecular Nutrition & Food Research*, Vol.53, pp. 625-634, ISSN 1613-4133.

Keep, E. (1975). Currants and gooseberries, In: *Advances in fruit breeding*, J. Janic & J. N. Moore, (Eds), 197-268, Purdue University Press.

Koponen, J.M.; Happonen, A.M., Mattila, P.H. Törrönen, A.R. (2007). Contents of anthocyanins and ellagitannins in selected foods consumed in Finland. *Journal of Agricultural and Food Chemistry*, Vol.55, pp. 1612-1619, ISSN 0021-8561.

Lee, J.; Finn, C E. (2007). Anthocyanins and other polyphenolics in American elderberry (*Sambucus canadensis*) and European elderberry (*S. nigra*) cultivars. *Journal of the Science of Food and Agriculture*, Vol.87, pp. 2665-2675, ISSN 1097-0010.

Li, T.S.C. (2002). Product Development of Sea Buckthorn, In: *Trends in new crops and new uses*, J. Janick and A. Whipkey, (Eds.), ASHS Press, Alexandria, VA.

Li, T.S.C.; Schroeder, W.R. (1996). Sea buckthorn (*Hippophae rhamnoides*): a multipurpose plant. *HortTechnology*, Vol.6, pp. 370-380, ISSN 1063-0198.

Li, W.; Hydamaka, A.W., Lowry, L., Beta, T. 2009. Comparison of antioxidant capacity and phenolic compounds of berries, chokecherry and seabuckthorn. *Central European Journal of Biology,* Vol.4, pp. 499-506, ISSN 1895-104X.

Liobikas, J.; Liegiūtė, S., Majienė, D. Trumbeckaitė, S., Bendokas, V., Kopustinskienė, D.M., Šikšnianas, T., Anisimovienė, N. (2008). Antimicrobial activity of anthocyanin-rich extracts from berries of diferent Ribes nigrum varieties. *Sodininkystė ir daržininkystė,* T. 27, pp. 59–66, ISSN 0236-4212 (in Lithuanian).

Lister, C.; Wilson, P., Sutton, K., Morrison, S. (2002). Understanding the health benefits of blackcurrants. *Acta Horticulturae,* Vol.585, pp. 443-449, ISSN 0567-7572.

Lohachoompol, V.; Srzednicki, G., Craske, J. (2004). The changes of total anthocyanin in blueberries and their antioxidant effect after drying and freezing. *Journal of Biomedicine and Biotechnology,* Vol. 5, pp. 248-252, ISSN 11107243.

Lugasi, A.; Hóvári, J., Kádár, G., Dénes, F. (2011). Phenolics in raspberry, blackberry and currant cultivars grown in Hungary. *Acta Alimentaria,* Vol.40, pp. 52-64, ISSN 1588-2535.

Määttä-Riihinen, K.R.; Kamal-Eldin, A., Mattila, P.H., Gonzalez-Paramas, A.M., Törrönen, A.R. (2004a). Distribution and contents of phenolic compounds in eighteen Scandinavian berry species. *Journal of Agricultural and Food Chemistry,* Vol.52, pp. 4477-4486, ISSN 0021-8561.

Määttä-Riihinen, K.R.; Kamal-Eldin, A., Törrönen, A. R. (2004b). Identification and quantification of phenolic compounds in berries of *Fragaria* and *Rubus* species (family Rosaceae). *Journal of Agricultural and Food Chemistry,* Vol.52, pp. 6178-6187, ISSN 0021-8561.

Mage, F. (1993). Vegetative, generative and quality characteristics of four blackcurrant (*Ribes nigrum* L.) cultivars. *Norwegian Journal of Agricultural Science,* Vol.7, pp. 327-332, ISSN 0801-5341.

Manach, C.; Scalbert, A., Rémésy, C., and Jiménez, L. (2004). Polyphenols: food sources and bioavailability. *American Journal of Clinical Nutrition,* Vol.79, pp. 727-747, ISSN 1938-3207.

Matsumoto, H.; Nakamura, Y., Hirayama, M., Yoshiki, Y., Okubo, K. (2002). Antioxidant activity of black currant anthocyanins and their glycosides measured by chemiluminescence in neutral pH region and in human plasma. *Journal of Agricultural and Food Chemistry,* Vol.50, pp. 5034-5037, ISSN 0021-8561.

Mazza, G. (1993). Anthocyanins in edible plant parts: A qualitative and quantitative assessment, In: *Anthocyanins in Fruits, Vegetables and Grains,* Mazza G, Miniati E., (Eds.), 119-140, CRC Press, Boca Raton, FL, USA, ISBN 0849301726.

Melo, M. J.; Moncada, M.C., Pina, F. (2000). On the red colour of raspberry (*Rubus idaeus*). *Tetrahedron Letter,* Vol.41, pp. 1987-1991, ISSN 0040-4039.

Montero, T.; Molla, E., Esteban, R., Lopez-Andreu, F. (1996). Quality attributes of strawberry during ripening. *Scientia Horticulturae,* Vol.65, pp. 239- 250, ISSN 0304-4238.

Moyer, R.A.; Hummer, K.E., Finn, C.E., Frei, B., Wrolstad, R.E. (2002). Anthocyanins, phenolics, and antioxidant capacity in diverse small fruits: *Vaccinium, Rubus,* and *Ribes. Journal of Agricultural and Food Chemistry,* Vol.50, pp. 519-525, ISSN 0021-8561.

Mullen, W.; McGinn, J., Lean, M.E.J., MacLean, M.R., Gardner, P., Duthie, G.G., Yokot,a T., Crozier, A. (2002). Ellagitannins, flavonoids, and other phenolics in red raspberries and their contribution to antioxidant capacity and vasorelaxation properties. *Journal of Agricultural and Food Chemistry*, Vol.50, pp. 5191-5196, ISSN 0021-8561.

Mullen, W.; Yokota, T., Lean, M.E.J., Crozie,r A. (2003). Analysis of ellagitannins and conjugates of ellagic acid and quercetin in raspberry fruits by LC-MS^n. *Phytochemistry*, Vol.64, pp. 617-623, ISSN 0031-9422.

Müller, L.; Fröhlich, K., Böhm, V. (2011). Comparative antioxidant activities of carotenoids measured by ferric reducing antioxidant power (FRAP), ABTS bleaching assay (αTEAC), DPPH assay and peroxyl radical scavenging assay. *Food Chemistry*, Vol.129, pp. 139-148, ISSN 0308-8146.

Nakajima, J.; Tanaka, I., Seo, S., Yamazaki, M., Saito, K. (2004). LC/PDA/ESI-MS profiling and radical scavenging activity of anthocyanins in various berries. *Journal of Biomedicine and Biotechnology*, Vol.5, pp. 241–247, ISSN 11107251.

Nohynek, L.J.; Alakomi, H-L., Kähkönen, M.P., Heinonen, M., Helander, I.M., Oksman-Caldentey, K-M., Puupponen-Pimiä, R. H. (2006). Berry Phenolics: antimicrobial properties and mechanisms of action against severe human pathogens. *Nutrition and Cancer*, Vol.54, pp. 18-32, ISSN 1532-7914.

Orak, H.H. (2006). Total antioxidant activities, phenolics, anthocyanins, polyphenoloxidase activities and its correlation in some important red wine grape varietes which are grown in Turkey. *Electronic Journal of Polish Agricultural Universities*, Vol.9, pp.1–7, ISSN 1505-0297.

Oszmianski, J.; Sapis, J.C. (1988). Anthocyanins in fruits of *Aronia Melanocarpa* (chokeberry). *Journal of Food Science*, Vol.53, pp.1241-1242, ISSN 1750-3841.

Oszmianski, J.; Wojdylo, A. (2005). *Aronia melanocarpa* phenolics and their antioxidant activity. *European Food Research and Technology*, Vol.221, pp. 809-813, ISSN 1438-2385.

Pantelidis, G.E.; Vasilakakis, M., Manganaris, G.A. and Diamantidis, Gr. (2007). Antioxidant capacity, phenol, anthocyanin and ascorbic acid contents in raspberries, blackberries, red currants, gooseberries and Cornelian cherries. *Food Chemistry*, Vol.102, pp. 777-783, ISSN 0308-8146.

Parr, A.J.; Bolwell, G.P. (2000). Phenols in the plant and in man. The potential for possible nutritional enhancement of the diet by modifying the phenols content or profile. *Journal of the Science of Food and Agriculture*, Vol.80, pp. 985-1012, ISSN 1097-0010.

Pecho, L.; Takač, J., Cvopa, J. (1993). Nutrient matter contents in fresh and processed currant fruits. *Acta Horticulturae*, Vol.352, pp. 205-208, ISSN 0567-7572.

Plocharski, W.; Zbroszczyk, J., Lenartowicz, W. (1989). Aronia fruit (Aronia melanocarpa, Elliot) as a natural source of anthocyanin colourants. 2. The stability of the color of aronia juices and extracts. *Fruit Science Reports* (Skiernewice), Vol.16, pp. 40-50, ISSN 0137-1479.

Poiana, M-A.; Moigradean, D., Raba, D, Alda, L-M., Popa, M. (2010). The effect of long-term frozen storage on the nutraceutical compounds, antioxidant properties and color indices of different kinds of berries. *Journal of Food Agriculture and Environment*, Vol.8, pp. 54-58, ISSN 1459-0263.

Raffo, A.; Paoletti, F., Antonelli, M. (2004). Changes in sugar, organic acid, flavonol and carotenoid composition during ripening of berries of three seabuckthorn (*Hippophae rhamnoides* L.) cultivars. *European Food Research and Technology*, Vol.219, pp. 360-368, ISSN 1438-2385.

Ranjith, A.; Kumar, K.S., Venugopalan, V.V., Arumughan, C., Sawhney, R.C., Singh, V. (2006). Fatty acids, tocols, and carotenoids of three sea buckthorn (*Hippophae rhamnoides, H salcifolia, and H tibetana*) grown in Indian Himalayas. *Journal of the American Oil Chemists' Society*, Vol.83, pp 359–364, ISSN 1558-9331.

Rubinskiene, M.; Jasutiene, I., Venskutonis, P.R., Viskelis, P. (2005). HPLC determination of the composition and stability of blackcurrant anthocyanins. *Journal of Chromatographic Science*, Vol.43, pp. 478-482, ISSN 0021-9665.

Rubinskienė, M.; Viškelis, P. (2002). Accumulation of ascorbic acid and anthocyanins in berries of *Ribes nigrum*. *Botanica Lithuanica*, Vol.8, pp. 139-144, ISSN 1392-1665.

Santos-Buelga, C.; Scalbert, A. (2000). Proanthocyanidins and tannin-like compounds – nature, occurrence, dietary intake and effect on nutrition and health. *Journal of the Science of Food and Agriculture*, Vol.80, pp. 1094-1117, ISSN 1097-0010.

Šarkinas, A.; Jasutienė, I., Viškelis, P. (2005). Mėlynių ir spanguolių ekstraktų antimikrobinės savybės [Antimicrobical properties of blueberry, cranberry and blackcurrant extracts]. *Sodininkystė ir daržininkystė*, Vol.24, pp. 100–111, ISSN 0236-4212 (in Lithuanian).

Scott, K.J. (2001). Detection and measurement of carotenoids by UV/VIS spectrophotometry, In: *Current protocols in food analytical chemistry*, Wrolstad R.E., Acree T.E., An H., Decker E.A., Penner M.H., Reid D.S., et al., (Eds.), John Wiley & Sons Inc., New York, ISBN 9780471142911.

Shin, Y.; Ryu, J.A., Liu, R.H., Nockand, J.F., Watkins, C.B. (2008). Harvestmaturity, storage temperature and relative humidity affect fruit quality, antioxidant contents and activity, and inhibition of cell proliferation of strawberry fruit. *Postharvest Biology and Technology*, Vol.49, pp. 201–209, ISSN 0925-5214.

Slimestad, R.; Solheim, H. (2002). Anthocyanins from black currants (*Ribes nigrum* L.). *Journal of Agricultural and Food Chemistry*, Vol.50, pp. 3228-3231, ISSN 0021-8561.

Slimstad, R.; Torkangerpoll, K., Nateland, H.S., Johannessen, T., Giske, N.H. (2005). Flavonoids from black chokeberries, Aronia melanocarpa. *Journal of Food Composition and Anglysis*, Vol.18, pp. 61-68, ISSN 0889-1575.

Slinkard, K.; Singleton, V. L. (1977). Total phenol analysis: Automation and comparison with manual methods. *American Journal of Enology and Viticulture*, Vol.28, pp. 49-55, ISSN 0002-9254.

Stoner, G.D.; Wang, L-S., Casto, B.C. (2008). Laboratory and clinical studies of cancer chemoprevention by antioxidants in berries. *Carcinogenesis*, Vol.29, pp. 1665-1674, ISSN 1460-2180.

Suthanthangjai, W.; Kajda, P., Zabetakis, I. (2005). The effect of hydrostatic pressure on the anthocyanins of raspberry (*Rubus idaeus*). *Food Chemistry*, Vol.90, pp. 193-197, ISSN 0308-8146.

Szajdek, A.; Borowska, W.J. (2008). Bioactive compounds and health-promoting properties of berry fruits: a review. *Plant Foods for Human Nutrition*, Vol.63, pp. 147-156, ISSN 1573-9104.

Tabart, J.; Pincemail, J., Defraigne, J.O., Dommes, J. (2006). Antioxidant capacity of black currant varies with organ, season, and cultivar. *Journal of Agricultural and Food Chemistry*, Vol.54, pp. 6271–6276, ISSN 0021-8561.

Talavéra, S.; Felgines, C., Texier, O., Besson, C., Manach, C., Lamaison, J.-L., Rémésy, C. (2004). Anthocyanins are effciently absorbed from the small intestine in rats. *Journal of Nutrition*, Vol.134, pp. 2275-2279, ISSN 1541-6100.

Tang, X.; Tigerstedt, P.M.A. (2001). Variation of physical and chemical characters within an elite sea buckthorn (*Hippophae rhamnoides* L.) breeding population. *Scientia Horticulturae*, Vol.88, pp. 203-214, ISSN 0304-4238.

Tian, Q.; Giusti, M.M., Stoner, G.D., Schwartz, S.J. (2006). Characterization of a new anthocyanin in black raspberries (*Rubus occidentalis*) by liquid chromatography electrospray ionization tandem mass spectrometry. *Food Chemistry*, Vol.94, pp. 465-468, ISSN 0308-8146.

Tiitinen, K.M.; Hakala, M.A., Kallio, H.P. (2005). Quality components of sea buckthorn (*Hippophaë rhamnoides*) varieties. *Journal of Agricultural and Food Chemistry*, Vol.53, pp. 1692–1699, ISSN 0021-8561.

Trajkovski, V.; Jeppson, N. (1999). Domestication of sea buckthorn. *Botanica Lithuanica*, Vol.2, pp. 37-46, ISSN 1392-1665.

Trajkovski, V.; Strautina, S., Sasnauskas, A. (2000). New perspective hybrids in breeding of black currants. *Sodininkystė ir daržininkystė*, Vol.19, No.3, pp. 3-14, ISSN 0236-4212.

Túlio, A.Z.Jr.; Reese, R.N., Wyzgoski, F.J., Rinaldi, P.L., Fu, R., Scheerens, J.C., Miller, A.R. (2008). Cyanidin 3-rutinoside and cyaniding 3-xylosylrutinoside as primary phenolic antioxidants in black raspberry. *Journal of Agricultural and Food Chemistry*, Vol.56, pp. 1880-1888, ISSN 0021-8561.

Univer, T.; Jalakas, M., Kelt, K. (2004). Chemical composition of the fruits of sea buckthorn and how it changes during the harvest season in Estonia. *Journal of Fruit and Ornamental Plant Research*, Vol.12, pp. 339-344, ISSN 1231-0948.

Valcheva-Kuzmanova, S.; Borisova, P., Galunska, B., Krasnaliev, I., Belcheva, A. (2004). Hepatoprotective effect of the natural fruit juice from *Aronia melanocarpa* on carbon tetrachlorideinduced acute liver damage in rats. *Experimental and Toxicologic Pathology*, Vol.56, pp.195-201, ISSN 0940-2993.

Vekiari, S.A.; Gordon, M.H., García-Marcías, P., Labrinea, H. (2008). Extraction and determination of ellagic acid content in chestnut bark and fruit. *Food Chemistry*, Vol.110, pp.1007-1011, ISSN 0308-8146.

Viškelis, P.; Bobinaitė, R., Rubinskienė, M. (2007). Biochemical composition and antioxidant properties of raspberries during ripening. *Sodininkystė ir daržininkystė*, Vol.26, pp. 161–173, ISSN 0236-4212.

Viskelis, P.; Rubinskienė, M., Bobinaitė, R., Dambrauskienė, E. (2010). Bioactive compounds and antioxidant activity of small fruits in Lithuania. *Journal of Food, Agriculture and Environment*, Vol.8, No.3&4, pp. 259-263, ISSN 1459-0263.

Viškelis, P.; Rubinskienė, M., Bobinas, C. (2008). Evaluation of strawberry and black currant berries intended for freezing and the methods of their preparation. *Journal of Food, Agriculture and Environment*, Vol.6, No.3-4, pp. 151-154, ISSN 1459-0263.

Viškelis, P.; Rubinskienė, M., Jasutienė, I. (2001). Influence of black currant pigments on berry technological properties. *Sodininkystė ir daržininkystė*, Vol.20, pp. 229-239, ISSN 0236-4212.

Vrhovsek, U.; Giongo, L., Mattivi, F., Viola, R. (2008). A survey of ellagitannin content in raspberry and blackberry cultivars grown in Trentino (Italy). *European Food Research and Technology*, Vol. 226, pp. 817-824, ISSN 1438-2385.

Vrhovsek, U.; Palchetti, A., Reniero, F., Guillou, C., Masuero, D., Mattivi, F. (2006). Concentration and mean degree of polymerization of *Rubus* ellagitannins evaluated by optimized acid methanolysis. *Journal of Agricultural and Food Chemistry*, Vol.54, pp. 4469-4475, ISSN 0021-8561.

Wada, L.; Ou, B. (2002). Antioxidant activity and phenolic content of Oregon caneberries. *Journal of Agricultural and Food Chemistry*, Vol.50, pp. 3495-3500, ISSN 0021-8561.

Walther, E.; Schnell, S. (2009). Black chokeberry (*Aronia melanocarpa*) – a special crop fruit. *Zeitschrift für Arznei und Gewurzpflanzen*, Vol.14, pp. 179-182, ISSN 1431-9292.

Wang, H.; Cao, G. and Prior, R.L. (1996). Total antioxidant capacity of fruits. *Journal of Agricultural and Food Chemistry*, Vol.44, pp.701-705, ISSN 0021-8561.

Wang, H.; Cao, G., Prior, R. (1997). Oxygen radical absorbing capacity of anthocyanins. *Journal of Agricultural and Food Chemistry*, Vol.45, pp. 304-309, ISSN 0021-8561.

Wang, S.I.; Zheng, W. (2005). Preharvest application of methyl jasmonate increases fruit quality and antioxidant capacity in raspberries. *International Journal of Food Science & Technology*, Vol.40, pp. 187-195, ISSN 1365-2621.

Wang, S.Y.; Chenand, C.T., Wang, C.Y. (2009). The influence of light and maturity on fruit quality and flavonoid content of red raspberries. *Food Chemistry*, Vol.112, pp. 676-684, ISSN 0308-8146.

Wang, S.Y.; Lin, H. (2000). Antioxidant activity in fruits and leaves of blackberry, raspberry, and strawberry varies with cultivar and developmental stage. *Journal of Agricultural and Food Chemistry*, Vol.48, pp. 140-146, ISSN 0021-8561.

Weber, C.A.; Perkins-Veazie, P., Moore, P.P., Howard, L. (2008). Variability of antioxidant content in raspberry germplasm. *Acta Horticulturae*, Vol. 777, pp. 493-498, ISSN 0567-7572.

Wrolstad, R.E. (2000). Anthocyanins, In: *Natural Food Colorant*, F.J. Francis, G.J. Lauro, (Eds.), 237-252, Marcel Dekker, Inc., NY, ISBN 0-8247-0421-5.

Wu, X.; Beecher, G. R., Holdel, J. M., Haytowitz, D. B., Gebhardt, S. E., Prior, R. L. (2006). Concentration of anthocyanins in common foods in the United States and estimation of normal consumption. *Journal of Agricultural and Food Chemistry*, Vol.54, pp. 4069-4075, ISSN 0021-8561.

Wu, X.; Gu, L., Prior, R.L., McKay, S. (2004). Characterization of anthocyanins and proanthocyanidins in some cultivars of *Ribes*, *Aronia*, and *Sambucus* and their antioxidant capacity. *Journal of Agricultural and Food Chemistry*, Vol.52, pp. 7846-7856, ISSN 0021-8561.

Yang, B.R.; Kallio H.P. (2001). Fatty acid composition of lipids in sea buckthorn (*Hippophae rhamnoides* L.) berries of different origins. *Journal of Agricultural and Food Chemistry*, Vol.49, pp. 1939-1947, ISSN 0021-8561.

Yao, Y.; Tigerstedt, P.M.A. (1994). Genetic diversity in *Hippophae* L. and its use in plant breeding. *Euphytica*, Vol.77, pp. 165-169, ISSN 0014-2336.

Yao, Y.; Tigerstedt, P.M.A., Joy, P. (1992). Variation of vitamin C concentration and character correlation and between and within natural sea buckthorn (*Hippophae rhamnoides* L.) populations. *Acta Agriculturae Scandinavica, Section B-Soil and Plant Science*, Vol.42, pp. 12-17, ISSN 0906-4710.

Zadernowski, R.; Naczk, M., Amarowicz, R. (2003). Tocopherols in sea buckthorn (*Hippophae rhamnoides* L.) berry oil. *Journal of the American Oil Chemists' Society*, Vol.80, pp. 55-58, ISSN 1558-9331.

Zeb, A. (2004). Important therapeutic uses of sea buckthorn (Hippophae): A review. *Journal of Biological Sciences* , Vol.4, pp. 687-693, ISSN 1727-3048.

Żurawicz, E.; Pluta, S., Danek, J. (2000). Small fruit breeding at the research institute of pomology and floriculture in Skierniewice, Poland. *Acta Horticulturae*, Vol.538, pp. 457-461, ISSN 0567-7572.

Part 5

Urban Horticulture

6

Urban Horticulture and Community Economic Development of Lagging Regions

Albert Ayorinde Abegunde
Department of Urban and Regional Planning,
Obafemi Awolwo University, Ile Ife,
Nigeria

1. Introduction

One of the global challenges of this era is climate change, arising from global warming and depletion of oxone layer; originally caused by reduction of green spaces on the earth surface. This is because only18% of earth surface is cultivable or capable of growing plants (Omisore and Abegunde, 2000). The rest is occupied by seas, mountains and ice (Encarta, 2005).The little area capable of growing plants (agriculture, landscaping, horticulture and green conservation) is highly competed for by housing, industrial and road constructions and incessant environmental disasters like bush burning, flooding, deforestation and settlement expansion due to urbanization. In other dimension, promoting green space development in the new millennium calls for special concerns from professionals in environmental sciences. The global need for increasing green areas have and may not be easily met in lagging regions of the world where war, famine, environmental pestilence and low income and inadequate infrastructure are on the increase. (Frey, 2000; Amati, 2008). In another dimension, these have made residents living in regions where basic needs of life are not met to care less for their environment (Food and Agricultural Organisation of the United Nations, 2010). In other words, lagging regions of the world are saddled with green challenge and economic depression. Until there is a pro-poor means of green revolution, many people in the third world nations may be least interested in contributing significantly in the face of poverty and famine that are limiting their growth and development (Food and Agricultural organization {FAO}, 2010).

The practice of urban horticultural garden in third world cities to boost food and ornamental plants production, provide job opportunity and promote green space development may bridge these gaps. This is because, urban horticulture (UH) make use of available pieces of land in cities to raise gardens that can be economically productive while contributing to environmental greening. This is why Moustier (1999) perceived it as an intensive production of a range of vegetables; aromatic, medicinal, flowering and ornamental plants grown mainly in the city or at its close periphery where there is competition among land uses. Adejumo (2003), Ward (1992) and Moss-Eccordt (1973) also opined that horticulture provides a physical green condition with appealing outlook that promotes good health and enhances the economic and other social values of communities, particularly when such plants are grown for commercial purpose.

Paradoxically, the art of practicing urban horticulture in developing nations is an unpopular path towards meeting ecological challenge of community greening, economic problem of famine and poverty and poor urban aesthetics. This study is an empirical expression of these concerns, citing Lagos, Nigeria as an example. The chosen city fits such study in a number of ways. First, it is considered as one of the Africa's fastest growing cities and commercial nerve centre for its country (Aluko, 2010). The general structure of land use distribution in the study area shows that only 520 hectares (2.8%) of the total land area is given to open space. These include all urban land for recreation, parks and gardens, urban agricultural land, commercial and individual horticultural gardens and unused spaces (Oduwaye, 2006). This is far below the 8-10% of land area expected to be made available for green space in a residential setting. Lagos, an area with limited land due to its closeness to the Atlantic Ocean, is also chocked with housing development, heavy industries and automobiles. Despite all these limitations, little attention has been given to spatial distribution of green space by government in the city (Abegunde *et al*, 2009).

There is the need to understand the economic implication of urban horticultural development in the study area and beyond. The conduct of this kind of study at the start of a new decade in the twenty-first century is imperative when urban agriculture is receiving its popularity as means of restoring productive green belts and economic revival to world cities. In other words, the aesthetics of urban horticulture that serves as a source of agricultural production in a poor economy is of concern to urban community development planners. The aim of the study is therefore to critically examine the spatial extent and practice of urban horticulture towards economic development of cities in lagging regions of the world, citing Lagos, Nigeria as an example. This is with a view to establishing the contributions of such practice to the social and economic development of the residents and the built environment and by this develop a framework that could be of importance to simultaneously further community greening and economic development of other lagging regions of the world.

2. The literature

2.1 The concept of urban horticulture (UH)

Horticulture is the art of gardening or plant growing; in contrast to agronomy - the cultivation of field crops such as cereals and animal fodder, forestry - cultivation of trees and products related to them, or agriculture – the practice of farming. Urban horticulture (UH) can also be seen as intensive production of a range of vegetables; aromatic, medicinal, flowering and ornamental plants grown mainly in the city or at its close periphery where there is competition among land uses (Moustier, 1999). The origin of horticulture lies in the transition of human communities from nomadic hunter-gatherers to sedentary or semi-sedentary horticultural communities, cultivating a variety of crops on a small scale around dwellings. (Von-Hagen, 1957; McGee and Kruse, 1986). A characteristic of horticultural communities is that useful trees are often planted around the built environment or specially retained from the natural ecosystem. The significance of this in promoting healthy environment is found in the works of Ebenezer Howard (1902) and further explained by Moss-Eccordt (1973) and Ademola, (2002). Thus, the practice of horticulture plays a role in the development of healthy communities in three distinct ways. First, it provides a physical condition with appealing outlook. Second, it promotes good health as carbon related gasses generated in cities are utilised during plants' photosynthesis while oxygen that is useful for man is released as bye product. Third, plants generally enhance the economic and social values of the community (Ward, 1992; Adejumo, 2003). This chapter is more inclined to the latter importance, though not disassociated with the former. This is because many urban horticulturists contribute their quota to vegetable production. The sales of these vegetables and ornamental plants provide markets for both horticulturists and middlemen and women in the business. This is why it has been argued that solutions to poverty in cities of developing countries has multiple faces, of which horticultural practice is one (Weinberger and Lumpkin, 2007).

2.2 Economy, green space challenge and horticultural practice in third world nations

Early socio-economic problems of the third world nations have been linked with poverty (Madzingira, 1997). That could be why Khan (2001) noted that most of the one-fifth of the world's populations afflicted by abject poverty, earning less than on Dollar a day live in lagging regions. Although, poverty have reduced over the past 40 years, particularly in China, India, South East Asia and South Africa, with little or no progress recorded in sub-Saharan Africa {Department for International Development (DFID), 2004}.

Specifically, between 1981 and 2001, the percentage of the number of people living on less than a dollar per day globally fell from 40.4% to 21.1% despite the 1.5 billion people that were added to the world populations within the same period. Of interest here is the inverse relationship between agricultural production and poverty in the world. For instance, Warr (2001) noted that growth in agriculture in a number of South East Asian countries significantly reduced poverty. This is just as Gallup *et al* (1997) had earlier observed that every 1% growth in per capita agricultural Gross Domestic Product (GDP) led to 1.61% growth in the incomes of the poorest 20% of the population – much greater than the impact of similar increases in the manufacturing or service sectors. In other words, agricultural related activities, of which urban horticulture is a part; are generally central to world poverty reduction. As argued by Weinberger and Lumpkin (2007) that horticultural products are facing increasing domestic and international demand, widening market access and helping residents in lagging regions of the world who engaged in such to escape poverty through production and exchange of non-staple crops.

In the past, the development policymakers having observed the role of agriculture in poverty reduction, focusing on staple grains, especially rice and wheat production across the globe. Recent observations by the Consultative Group on International Agricultural Research (CGIAR) has expressed more interest in horticulture and research on high value crops as priorities (CGIAR, 2004), though investments in horticultural research and products remain inadequate. Despite this inadequacy, on a global scale, the value of all fruits and vegetables traded as horticultural products is more than double the value of all cereals traded as farm products (FAO, 2005). In addition, farmers in other regions of the world have also found it profitable to expand production of horticultural produce at the expense of the cereal area (Weinberger and Lumpkin (2007). This is not unconnected with the ease of practicing horticulture where land seems to be inadequate for extensive cropping. For instance, some residents engage in indoor and street horticulture, enclosed gardening, potted plants to mention but few. Of interest here are the multiple economic opportunities in horticultural practice in modern cities. This is despite its contribution to community greening in the global warming era.

As observed by Food and Agricultural Organisation (FAO), (2010), most rapidly growing cities in the world are located in developing countries of Asia and Africa, where rapid urbanisation is at variance with green space development (Thanh, 2007) and the practice of urban agriculture is at high demand. Thus, in such cities, horticultural gardening bridges the gap between poverty, environmental pollution arising from urbanization and reduction in green area (Abegunde, 2011). In other words, three out of the four targets of millennium development goal (poverty eradication, reduction in global warming, health and education for all) can be achieved through support of urban horticulture. Hence, researchers currently are unveiling the self-help pro-poor and environmental opportunities endowed in the practice {Moustier, 1999; Food and Agricultural organization (FAO), 2010}.

The concerns in this chapter are that horticulture is an easy to practice arm of agriculture that intensively utilizes little space of land, even in core areas of the built environment (Abegunde *et al*, 2009). Its ability to combine aesthetics through landscaping with production of food crops at a reasonable and manageable scale makes it a welcome practice both to the poor and the elites in combating poverty and enhances environmental beautification (Digkistra and Magori, 1995; Chweya, *et al*, 1995; Weinberger and Lumpkin, 2007). It is seen as one of the easy ways of increasing urban green space and by this, promotes good health (Abegunde, 2011). In another dimension, horticulture is an appropriate approach towards environmental friendly pro-poor development in the global warming era. The premise of this chapter is that though the practice of urban horticulture in developing nations is an unpopular path (towards meeting ecological challenge of community greening, poverty alleviation and urban aesthetics), it can contribute towards solving part of the environ-economic problems of the world.

2.3 Green economy theory

In recent years, poverty and environmental issues have been attracting significant attentions in development studies (Nyasha, 1997; Frey, 2000). First, the causes and consequences of poverty have been explored and theoretical models have been developed to explain hitherto obscure causalities. Along this line, scholars who are environmentally oriented have been attempting to create a meeting point for these two concerns, developed environmentally oriented theories that are pro-poor in approaches (Amati, 2008).

Among such theories is green economy model which focused on improved human well-being and social equity, while significantly reducing environmental risks and ecological scarcities. It is a new theory built on the platform of pro-growth model developed lately to justify the need for the type of development that is low income friendly and poverty eradicating. Origin of this can be traced to the work of Simon Kuznets, who in 1955 found an inverted-U pattern between per capita income and inequality based on a cross-section of countries. According to him, as per capita income rises, inequality first worsens and then improves. The major driving force was presumed to be structural change that occurred because of labor shifts from a poor and less productive traditional sector to a more productive and differentiated modern sector (Kakwani, Khandker and Son, 2004). This was latter supported by Kravis (1960), Oshima (1962), Adelman and Morris (1971), Paukert (1973), Ahluwalia (1974, 1976), Robinson (1976), and Ram (1988). Recent writings on this can also be found in the works of Anand and Kanbur (1984), Fields (1989), Oshima (1994), Deininger and Squire (1996) and Ravallion (2001). The common arguments in their works are that there is a strong and complex relation between growth and poverty and these are determined by the level and changes in inequality.

Pro-poor growth is therefore concerned with the interrelationship between these three elements: growth, poverty, and inequality. In addition, it is the kind of growth that benefits the poor and provides them with opportunities to improve their economic situations (UN 2000, OECD 2001). Linking this to green economy model, green pro-poor economy of growth is one whose growth in income and employment is driven by public and private investments in green related programmes that reduce carbon emissions and pollution, enhance energy and resource efficiency, and prevent the loss of biodiversity and ecosystem services. This development path should maintain, enhance and, where necessary, rebuild natural capital as a critical economic asset and source of public benefits, especially for poor people whose livelihoods and security depend strongly on nature.

In relation to this study, green economy is an attempt at accomplishing social, ecological and economic development of lagging regions through low carbon, resource efficiency and social inclusion. A green economy as a new paradigm is the one that believes that sustainability can be achieved through recognition of the cardinal roles and combine efforts of green aspect of the environment and community economy that alleviates poverty and improves green areas, turning lagging regions to prosperous ones. It is a new model of growth that is much less intensive in natural resources and that can lead to social well-being and poverty reduction in Africa and beyond. It opines that the simple pathway towards sustainable development is to balance and coordinate different interests: between economic growth/job creation and environmental integrity, between the rich and the poor, and between the present and the future generations.

As a new model in the green environment, it aims at achieving millennium development goals through pro-growth, pro-jobs and pro-poor techniques of turning environmental imperatives into viable economic activities, helps reconcile the need for economic growth and the need to ensure the environmental basis for continued growth into the future. It recognizes the role of green industry in economic transformation. Green industry here refers to but not limited to businesses involved in production, distribution and services associated with ornamental plants, landscape and garden supplies and equipment. Segments of the industry also include wholesale nursery, greenhouse and sod growers, landscape architects, contractors and maintenance firms on green issues, retail garden centres, home centers and mass merchandisers with lawn and garden departments, and marketing intermediaries such as brokers, horticultural distribution centres, and re-wholesalers. In addition to these are commercial sectors, many state and local governments' institutions that are related to urban forestry operations for management of parks, botanic gardens, and right-of-ways. The Green Industry is linked to urban forestry, by providing quality plant material and professional personnel with specialized expertise for growing, maintaining, and managing city trees. Horticulture as mentioned under the green industry is the science and art involved in the cultivation, propagation, processing and marketing of ornamental plants, flowers, turf, vegetables, fruits, and nuts. Within the horticultural sector, the environmental horticulture industry, often referred to as the "Green Industry", is one of the fastest growing sectors of agriculture in the US (Palma and Hall, 2009). This chapter therefore sees green economy model as pro-poor theory of sustainability through involvement in horticultural practice to improve residents' income and boost national economic development while paying attention to community greening, urban aesthetics, and ecological balance in global warming era.

3. Research methodology adopted

This study focused on Lagos, Nigeria and used information generated from structured questionnaire administered on residents who engaged in outdoor commercial horticultural practice along major streets in Eti-Osa Local Government Area of the city. Reconnaissance survey revealed that urban commercial horticulture in Lagos city was practiced by private individuals and most of their gardens were not formally planned or located in government designated places. The study purposively selected all (100%) the existing seventy-five (75) gardens and targeted their owners (managers) during questionnaire administration. The study administered questionnaire on all the identified practitioners of urban commercial horticulture in the area. Sixty-three (63) of them were eventually interviewed, as the managers (owners) of the remaining 12 gardens were not available for questionnaire administration. Analysis in this study was based on these 63 respondents. Questions asked the respondents centred on their economic background, amount invested in the horticultural gardens, cost of production and profit realized annually, number of workers engaged in the practice and amount of money paid to them annually. Others are the horticulturists' social and economic contributions to the development of their community among others. Data for the study were analysed using descriptive and inferential statistics. Specifically, the study employed frequency tables to analyse data collected on area of land available for the practice, the economic characteristic of the horticulturists. Regression model was employed to confirm the significance of variables used to measure the economic contributions of the urban horticulture and profit realized by the managers in Lagos, Nigeria.

4. Results and discussion

4.1 Economic background of commercial horticulturists in Lagos, Nigeria

Table 1 reflects the economic characteristics of commercial horticulturists in the study area. It shows that more than two-third (63.6%) of them earned above N29,999 (Nigeria currency) per month (about 6.25 US Dollars per day). Specifically, one-quarter (25%) and less than one-sixth of them earned above N49,999 (10 US Dollars) and below N20,000 (4 US Dollars) monthly as profit realized from horticultural practice respectively. This is a clear indication that residents who were into commercial horticulture in Lagos, Nigeria were well living above the poverty line of income below 1.25 US Dollars per day (Lustig and McLeod, 1997; Kubelková, 2007). This is significant in an economy of a developing nation like that of Nigeria.

Income (in Nigeria Naira)	Frequency	Percentage
Below 20,000	10	15.9
20,000-29,999	13	20.4
30,000-39,999	16	25.4
40,000-49,999	09	14.3
50,000-59,999	06	9.6
Above 60,000	09	14.4
Total	63	100

Source: Author's field survey data, 2009.

Table 1. Monthly Income of Urban Commercial Horticulturists in Lagos, Nigeria

4.2 Economic contribution of urban horticulture in Lagos, Nigeria, through employment and community greening

A regression analysis of the significance of the urban commercial horticulture through profits realized by the managers on other independent variables shows that four out of the tested five variables in Table 2 were significant with R^2 = 50.5%, $F_{4,58}$ = 16.813 at p<0.005. These predictor variables are number of staff employed in horticultural practice (p=0.001), money paid to workers engaged in the practice annually (p=0.000), money spent on trees planted in the community for public benefit (p=0.000) and money given out for community development in Lagos (p=0.000). Only money paid annually to government purse as tax was not significant in the test (0.145). The latter may indicate that the horticulturists were not committed to their payment of government tax in the study area. In the contrary, the former might have also indicated that commercial horticultural practice in Lagos, Nigeria, though an informal sector; have contributed significantly to the economy in a measure by providing job for some workers and paying their wages/salaries. In addition, there are indications that their economic contributions to the community development programmes through donation of cash and expending money in planting trees for public use in the study area were significant due to profit realized from the practice.

Predictor	Beta	P
Number of staff employed in horticultural practice	-1.203	.001
Money paid to workers engaged in the practice annually	1.628	.000
Money paid annually to government purse as tax	0.188	.145
Money spent on trees planted in the community for public benefit	0.228	.003
Money given out for community development	0.157	.000

(Only Money paid annually to government purse as tax was not significant in the analysis)
The adjusted $R2$ = 50.5%, F4,58 = 16.813, p<0.005.

Table 2. Regression Analysis between Profits realized through Urban Commercial Horticultural Practice and other Predictor (Economic) Variables in Lagos State, Nigeria.

Further examination of the socio-economic impact of the studied horticulturists was revealed in their contributions through greening of their community. Regression tests conducted on this, using predicting variables on horticultural community greening through planting of trees (p=0.000), flowering plants (p=0.003) and fruit trees (p=0.379) for public consumption revealed that the former two were statistically significant (at p < 0.005) among the tested variables. This is an indication that urban horticulture in the study area contributed to the social development of their community. This is because tree planting improves aesthetics and reduces carbon content in the atmospheric air of the environment. This is imperative in the global warming era. The inability of the urban horticulture in the area to contribute significantly to planting fruit trees might have been due to lack of adequate space for the practice, as reconnaissance survey revealed that most of them utilized road setbacks for the practice as seen in Plate 1.

Predicting Variables on plants planted by Horticulturists in public places	Beta	P
*Number of trees nursed and panted in public places for community benefit	0.545	.000
*Number of flowering plants and hedges nursed and panted in public places for community's benefit	0..438	.003
Number of fruit trees/crop plants planted for public consumption	-0.358	.379

(Only fruit/crop trees planted for public use was not significant in the test)
*Significant variables at $p<0.005$. The adjusted R^2 = 32.20%, $F_{2,60}$ =6.891

Table 3. Regression Analysis predicting Relationships between Socio-economic Impact of Commercial Horticultural Practice and other Independent Variables in Lagos State, Nigeria.

Plate 1. One of the Commercial Horticulture Gardens, together with Trees planted for Public use along the Street in Eti-Osa Local Government Area, Lagos State, Nigeria.

4.3 Land area covered by urban commercial horticultural gardens in Lagos, Nigeria

Land area used for urban commercial horticulture for this study was obtained from data collected through physical measurement of each plot of land used for the practice and information obtained from secondary source on total area covered by land and water bodies in the study area. Below is a simple mathematical calculation of the land area used for horticultural gardening in the study area:

Total area of land (and water bodies) in Eti=Osa Local Government{y} =**3060175 m²**
(As documented in Abegunde, 2011)
Total area covered by water bodies   {b}   =   758903 m²
(As documented in Abegunde, 2011)
Total land area (excluding area used for street horticultural gardens in the study area) {c};
where 'c' is to be computed.
Total land area devoted to street horticultural gardens {a}; where 'a' is to be computed.
If a = $(g_1 + g_2 + g_3 + g_3 + g_4 + ... g_{63})$ *as documented in Abegunde (2011), and validated through physical measurement of each area of land in the sampled horticultural gardens in the study area during reconnaissance survey.*

Where g = plot size of each street horticultural garden (m²)
And $(g_1 + g_2 + g_3 + g_3 + g_4 + ... g_{63})$ = all the sixty-three (63) existing and sampled horticultural gardens in the study area

a = total land area devoted to street horticultural gardens (m²)
a = (120 + 128 + 159 + 168 + 168 + 170 + 180 + 190 + 202 + 210 + 210 + 240 + 310 + 338 + 344 + 350 + 350 + 355 + 363 + 386 + 396 + 400 + 431 + 445 + 471 + 483 + 487 + 495 + 496 + 498 + 500 + 503 + 510 + 518 + 520 + 528 + 530 + 538 + 539 + 554 + 555 + 560 + 562 + 577 + 584 + 590 + 602 + 612 + 612 + 635 + 640 + 668 + 678 + 798 + 820 + 837 + 842 + 870 + 871 + 899 + 900 + 980 + 1200)m² = **31675.** Then, a = **31675 m².**
If   y = a + b + c,
y - (a+b) = c
3060175 – (31675 + 758903)=   c
3060175 – 790587   =   c
C =   2269588 m²   or   560.83 Acres {Total land area in Eti-osa Local Government (excluding areas used for street horticultural gardens}
% of land use for commercial horticulture garden

$$= \quad \frac{a}{c} \times \frac{100}{1} \quad = \quad \frac{31675}{2269588} \times \frac{100}{1} = \quad 1.4\%$$

The study shows that the total area used for commercial horticultural gardens alone, which contributed to urban green space is 1.4% of the entire study area. Although this percentage may exclude the area formally planned for green space in Eti-osa Local Government, Lagos, Nigeria because most of them were practiced on road setbacks. However, they have added to community greening of the environment.

5. Concluding remarks

The study revealed that urban horticulture can contribute to economic development of residents in developing nations and beyond. This is evident in the daily earnings accrued to those engaged in the practice. In addition to this are their contributions to the community's economy through employment of staff and payment of their wages/salaries, economic contribution to community development and planting of trees and flowering plants for public use in the environment. Although the area of land used for urban commercial horticultural practice in Lagos, Nigeria seemed to be less than 1.5% of the total land area, its economic advantage is a justification that there is the need to look inward on commercial horticulture in urban areas of lagging regions of the world to cope with global economic recession.

6. Acknowledgment

The author wants to acknowledge the relentless efforts of Alida Lesnjakovic and Sasa Leporic in making sure that this manuscript was completed and submitted. Without their efforts, this chapter would never be contributed to this book. The special concern of In-Tech Publishers for articles that scientifically uplift knowledge and willingness to air the views of scholars from lagging regions of the world at no cost are highly appreciated.

7. References

Abegunde, A.A. (2011). Community approach to growing greener cities through self-help street horticultural gardens: a case study of Lagos, Nigeria. *British Journal of Environment & Climate Change 1(3): 103-117, 2011* SCIENCEDOMAIN *international www.sciencedomain.org*

Abegunde, A.A., Omisore, E.O., Oluodo, F., Olaleye, D. (2009). Commercial horticultural practice in Nigeria; its socio-spatial effects in Lagos City. *African Journal of Agricultural. Resource, 4(10). www.academicjournals.org*

Adelman, I. and C.T. Morris. (1971) *Economic growth and social equity in developing countries.* Stanford, CA: Stanford University Press

Adejumo RO (2003). Development strategy for sustainable public park system in metropolitan Lagos. *The City in Nig.* Obafemi Awolowo University Press, Ile-Ife. Pp. 112- 120.

Ademola A (2002). Urban art and aesthetics in Nigeria. *The City in Nig.* Obafemi Awolowo University Press, Ile-Ife. Pp. 212-218

Ahluwalia, M.S. (1974) Income inequality: some dimensions of the problem. In H.Chenery et al. (Edts.), *Redistribution with growth.* Oxford: Oxford University Press

Ahluwalia, M.S. (1976) Inequality, poverty and development. *Journal of Development Economics,* Vol. 3(4): 307-42.

Aluko, O.E (2010). The Impact of urbanization on housing development: The Lagos Experience, Nigeria. Ethiopian J. Environmental. Studies. and Management., 3(3).

Amati, Marco (2008). *Urban green belts in the twenty-first century.* Ashgate Publishing, Ltd

Anand, S. and Kanbur, S.M.R. (1984) The Kuznets process and the inequality development relationship, *Journal of Development Economics,* Vol. 40, 25-52

CGIAR (2004). Revised Summary Report on Developing CGIAR System Priorities for Research. Science Council working paper. Washington DC: CGIAR

Deininger, K. and L. Squire (1996) "Measuring income inequality: a new data-base" *World Bank Economic Review,* Vol. 10(3): 565-91

DFID Agriculture, Growth and Poverty Reduction Agriculture and Natural Resources Team of the UK Department for International Development (DFID) in October 2004Rural Poverty in Developing Countries

Duclos, J. and Wodon, Q. (2003). "Pro-Poor Growth". Unpublished mimeo. World Bank: Washington DC

Encarta (2005). Encarta on World Population and Land Coverage. Microsoft Encarta Reference Library, 2005.

FAO (2005). FAOSTAT data. Accessed November 2004, http://www.fao.org.

Fields, G.S. (1989) "Changes in poverty and inequality in the developing countries" mimeo. Food and Agricultural Organisation (FAO) (2010). Growing greener cities. FAO Publishing Policy and Sppt. Rome, Italy

Frey, H.W. (2000). *Not green belts but green wedges: the precarious relationship between city and country*. Urban Design International, Stockton Press. www.stockton-press.co.uk/udi

Gallup, J., S. Radelet and A. Warner (1997). Economic growth and the income of the poor.CAER Discussion Paper No. 36. Harvard Institute for International Development: Cambridge, MA, USA.

Howard, Ebenezer (1902). *Garden Cities of Tomorrow*. London: S. Sonnenschein & Co., Ltd.

Jana Kubelková, (2007). (BSc. Thesis) Charles University in Prague Natural Science Faculty, Department of Social Geography and Regional Development Kakwani, N. (1980) "On a Class of Poverty Measures" *Econometrica*, Vol. 48, No. 2, pp 437-446

Katinka Weinberger and Thomas A. Lumpkin, (2007). Diversification into Horticulture and Poverty Reduction: A Research Agenda The World Vegetable Center, Shanhua, Taiwan

Kravis, I.B. (1960) "International Differences in the Distribution of Income" *Review of Economics and Statistics*, Vol. 42: 408-16

Kuznets, S. (1955) "Economic Growth and Income Inequality", *American Economic Review*, Vol. 45, 1-28

Lustig, Nora Claudia and McLeod, Darryl (1997): Minimum wages and poverty in developing countries: some empirical evidence. In Edwards, Sebastian and Nora Lustig (Edits.), *Labor Markets in Latin America* (Brookrings Institution Press, Washington DC)

Madzingira, Nyasha (1997). Poverty and Ageing in Zimbabwe. *Journal of Social Development in Africa* (1997). 12.2,5-19

Mahmood Hasan Khan, (2001). Implications for Public Policy International Monetary Fund, March 2001

Marco A. Palma, and Charles R. Hall (2009). The Economic Impact of the Green Industry in Texas. Summary Report to the Texas Nursery and Landscape Association 2009. Texas AgriLife Extension Service Texas A&M University System.

McGee, J.R. and Kruse, M. (1986): Definitions of Horticulture. Online Botanical Encyclopedia. (http://www.eplantscience.com/botanical_biotechnology_biology_chemistry/horticulture.php).

Moss-Eccordt (1973). An Illustrated Life of Sir Ebenezer Howard, 1950-1928. Shire Publications Ltd. United Kingdom, 10pp.

Moustier, P. (1999). Définitions et contours de l'agriculture périurbaine en Afrique Subsaharienne. In: P. Moustier, A. Mbaye, H. de Bon, H. Guérin, J. Pagès (eds), Agric. périurbaine en Afrique subsaharienne, CIRAD, Colloques, pp. 17-29.

Nanak Kakwani, Shahid Khandker, Hyun H. Son, (2004), Pro-poor growth: concepts and measurement with country case studies. International Poverty Centre, United Nations Development Programme

Oduwaye L (2006). Effects of globalization on Lagos cityscape *Res. Rev. Ns 22.2* (2006) 37-54

OECD (2001). Rising to the global challenge: partnership for reducing world poverty. Statement by the DAC High Level Meeting. April 25-26., 2001. Paris: OECD

Omisore, E. O. and Abegunde, A. A. (2000). Land management and environmental degradation: some preventive strategies in the new millennium. *Journal of the Nigerian Anthropological and Sociological Association*, (A Publication of the Nigerian Anthropological and Sociological Association) Ile-Ife, pp 124-136.

Oshima, H. (1962). International comparison of size distribution of family incomeswith special reference to Asia. *Review of Economics and Statistics*, Vol. 44: 439-45

Oshima, H. (1994). Kuznets curve and Asian income distribution. In T. Mizoguchi(ed.), *Making Economies More Efficient and More Equitable: Factors DeterminingIncome Distribution*. Economic Research Series, No. 29. The Institute of EconomicResearch, Hitotsubashu University. Tokyo: Oxford University Press

Paukert, F. (1973). Income distribution at different levels of development: a surveyof evidence. *International Labour Review*, Vol. 108(2): 97-125

Ram, R. (1988). Economic development and income inequality: further evidenceon the u-curve hypothesis. *World Development, Vol. 16(11)*: 1371-1376

Ravallion, M. (2001). Growth, inequality and poverty: looking beyond averages. World Development, 29-11, 1803-1815 22 International Poverty Centre Working Paper no° 1

Robinson, S. (1976). Sources of growth in less developed countries. *Quarterly Journal of Economics* , Vol. 85(3): 391-408

Son, H. (2004). Pro-poor growth: definitions and measurements. Unpublished memo. World Bank: Washington D.C.

Thanh, L. (2007). Economic development, urbanization and environmental changes in Ho Chi Minh City, Vietnam: Relations and policies. Paper presented to the PRIPODE workshop on; Urban *Pop., Devpt. and Env. Dynamics in Developing. Countries.* Nairobi, Kenya

Von Hagen, V.W. (1957) Lectures 14, 15, and 16 Horticulture of Pre-Columbian America (http://www.hort.purdue.edu/newcrop/Hort_306/text/lec14.pdf)

Ward S.V. (1992). *The Garden City; past present and future*. E and FN Spon., London. 2.

Warr, P. (2001) *Poverty reduction and sectoral growth, results from South East Asia.*Australia National University: Canberra, Australia.

7

Local Botanical Knowledge and Agrobiodiversity: Homegardens at Rural and Periurban Contexts in Argentina

María Lelia Pochettino, Julio A. Hurrell and Verónica S. Lema
Laboratorio de Etnobotánica y Botánica Aplicada,
Facultad de Ciencias Naturales y Museo,
Universidad Nacional de La Plata,
Consejo Nacional de Investigaciones Científicas y Técnicas (CONICET),
República Argentina

1. Introduction

1.1 Ethnobotany and horticulture

Homegardens are defined as those cultivated spaces, generally of reduced extensión, located in the surroundings of the house. Garden produce is mainly consumed at home, or given away to related families, but exceptionally devoted to commercialization as a supplementary resource of domestic economy (Buet *et al.*, 2010; Pochettino, 2010, Wagner, 2002). Homegardens study constitutes a subject of increasing interest in Ethnobotany, as this approach contributes to both agrobiodiversity conservation (in particular to the infraspecific level) and to the preservation of cultural diversity: management strategies as well as species and varieties selection are not market-oriented, but they are regulated by preferences and culinary uses, linked with family traditions. So, these homegardens could be considered as real adaptative responses of local human groups arising from their own experience in the environment. This subject has been approached by diverse authors all over the world (Albuquerque *et al.*, 2005; Blanckaert *et al.*, 2004; Das & Das, 2005; Lamont *et al.*, 1999; Nazarea, 1998; Vogl *et al.*, 2002; Vogl *et al.*, 2004; Vogl-Lukasser *et al.*, 2002; Wagner, 2002; Watson & Eyzaguirre, 2002) even in Argentina, many of them developed by the research team of Laboratorio de Etnobotánica y Botánica Aplicada (LEBA), Facultad de Ciencias Naturales y Museo, Universidad Nacional de La Plata, Argentina (Buet *et al.*, 2010; Del Río *et al.*, 2007; Lema, 2006; Maidana *et al.*, 2005; Martínez *et al.*, 2003; Pérez *et al.*, 2008; Pochettino *et al.*, 2006; Pochettino, 2010; Turco *et al.*, 2006).

The context for the study of horticultural practices performed in homegardens is centered on the botanical knowledge that guides those practices. *Botanical knowledge* (BK) is a set of knowledge and beliefs about plant environment, that conditions the strategies of plant selection and handling, specifying wich plant should be considered as a resource, and how it should be managed. BK shows different features according with cultural and ecological conditions. In many areas, horticultural practices derive from a local BK characteristic of culturally contexts generally viewed as homogeneous, mainly because a long experience of human group in its environment, where knowledge is transmitted from generation to generation orally and in the shared practices. This BK has been named *traditional* (TBK) and there has been a rise in the number of studies about TBK because they are usually endangered and their rescue is urgent (Balick & Cox, 1996; Castellano, 2000; Gadgil *et al.*, 1993; Maffi, 2001; Pochettino & Lema, 2008). In periurban zones surrounding large urban areas (conurbations) — and that are a sort of dynamic ecotone between these and rural areas, (Barsky, 2010)— local BK influencing horticulture does not constitute a TBK in the sense defined above, because periurban areas are immerse in an heterogeneous cultural context, with human groups without a long experience in the environment (in comparative terms), and where knowledge is transmitted through social means of communication besides oral ones.

The ethnobotanical research on homegardens carried out by our team, have been adressed to two cases that are representative of both contexts described above. The first one is a rural context (culturally homogeneus) in Northwestern Argentina, where horticulture is practiced since 2000 years ago and strong local traditions can be traced. The second one is a periurban context (culturally heterogeneus) in Buenos Aires province, where settlements are no older than 130 years and the relationships human beings-environment are the result of the combination of factors of diverse origin, linked to traditions or non traditional influences. In both cases, the most evident difference is temporal depth in the relationship human beings-plant environment. Nevertheless, other parameters have also been analyzed in order to characterize local BK: both sociocultural and environmental contexts. For the purposes of analysi swere considered homegardens features, cultivated species and varieties, their origin, management practices and selection criteria applied to them and related to local BK and values.

1.2 Study areas

1.2.1 Northwest of Argentina

The Northwest of Argentina (NOA) (Fig. 1) outstands because of a notorious biodiversity all along its extensive geography, and because of an equally rich cultural diversity as the result of a social development deeply rooted in ancient times. Consequently this is one of the areas of Argentina that shows the greater biocultural diversity. Landscape is characterized by mountain chains connected by deep fluvial valleys. As a result, a network of pathways linking different ecological zones — from eastern lowlands to dry uplands— was established since the first settlements in the area 10.000 years ago in order to allow exchange and complementary use of such diverse environments (Nielsen, 1996; Rodríguez & Aschero, 2007; Tarragó, 1980). Between 3000 BP and 1000 AD local processes of camelids domestication and large-scale environmental modifications took place, related with the establishment of the first farmer's villages, in which the control of production was handled by domestic units in all altitudinal zones (lowlands, valleys and high plateau or *Puna*) (Albeck, 2003/2005; Berberián & Nielsen, 1988; Lema, 2009; Quesada, 2006; Yacobaccio, 2007). In the *Puna* it has been stated that the beginnings of plant domestication and the generation of productive spaces did not imply the loss of those practices linked to gathering of wild plants, as well as the tolerance of those species and other weeds in the areas devoted to gardening, thus generating wild-weedy-crop complexes (Lema 2009). In the course of cultural development of NOA, productive zones were increased and enlarged, and a technology and architecture typical of the area were developed, characterized by the use of *pircas* (stone fences of prehispanic origin) to delimit irrigation channels, cultivation fields and terraces in mountain slopes (Berberian & Nielsen 1988, Nielsen 1996). Local cultivation modes, adapted to environmental particularities of the different ecological zones, were developed, for instance cultivation fields with high walls in the *Puna*, where they acted as heat regulators in a zone with low temperatures and marked daily thermal amplitude (Albeck, 2003/2005; Albeck *et al.*, 2008). This productive development was accompanied by changes in production management, which fell into local communitarian control between 1000 and 1470 AD to be then enclosed in the sphere of Inka control between 1471 and 1536 (Capparelli *et al.*, 2011).

After European Conquest indigenous spaces were disrupted, and the forced establishment of a new regime of exploitation and production resulted in drastical changes, as the introduction of exotic cattle and new crops like wheat, barley and peaches (Capparelli *et al.*, 2005), along with the changes in the management of productive areas and their space configuration (Boixados, 2002; Tarragó, 2000). At present, the area is inhabited by creoles and natives that are descendents of different etnias. They possess a regime of land exploitation of rural kind at low scale. In spite of those processes of intense change and colonization in the area, it is remarkable that diverse models of exploitations and transformations of the environment, oriented by TBK within local communities, still co-exist nowadays, in the NOA.

Field works have been performed in 4 peasant communities in the NOA: El Shincal, in the Valley of Hualfin, Catamarca province; Rachaite and Coranzulí, in the *Puna* of Jujuy province, and Santa Victoria Oeste in Salta province, in the upper basin of Bermejo river. Peasant community of El Shincal is settled 7 km far away from the small villaje of Londres de Quimivil in the Department of Belén, Catamarca. From the geographical point of view it is located in the Pipanaco basin and belongs to the *Monte* phytogeographical province, which is characterized by a dry shruby vegetation, being its warm climate, dry continental and subtropical (Capparelli *et al.*, 2004). Argentine *Puna* constitutes a high plain that comprises part of Jujuy, Salta and Catamarca provinces, being accessed from eastern longitudinal valleys by three main ravines: Humahuaca, del Toro and the northern sector of Calchaquíes Valleys (Merlino & Rabey, 1978). In general, *Puna* is characterized as an herbaceous-shruby stepe with meadows or *ciénagos* with permanent water in the lower areas (Cabrera, 1976). Coranzuli and Rachaite are two small communities of Jujuy province located in high ravines of the Guayatayoc basin at *ca.*3800 meters over the sea level. Santa Victoria Oeste is the capital of the Department of Santa Victoria, in the northern extreme of Salta province. This district is bounded by Bolivia to the north, by Orán and Iruya departments to the south and east and by Jujuy province to the west. Santa Victoria Oeste is settled, at 2500 m over the sea level, in a valley with high grazing lands belonging to the phytogeographical province of *Yungas* (Cabrera, 1976), *ca.* 2500-4000 m over sea leavel. Over this ecological floor it is the typical vegetation of high Andean mountains, and below it, cloud forests and woods typical of *Yungas*. Ecological features of the fertile valleys where high grazing lands are found, conditioned not only human settlements but also the development of a diverse horticultural activity (Hilgert & Gil, 2008; Hurrell, 1990, 1991; Hurrell & de la Sota, 1996).

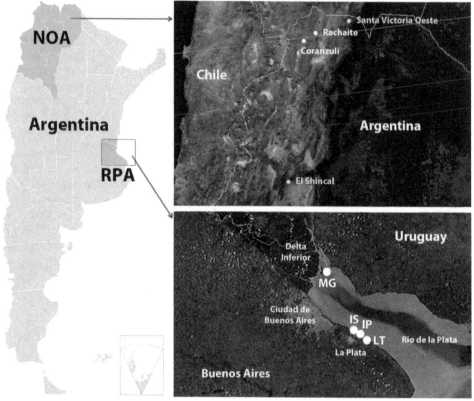

Fig. 1. Study areas: NOA, Northwest of Argentina area; RPA, Río de la Plata area. Sampling locations in both areas on satellite image of NASA. MG: Isla Martín García; IS: isla Santiago; IP: Isla Paulino; LT: Los Talas

1.2.2 Río de la Plata area

Several homegardens are located in the coastal area of the northeast of Buenos Aires province, called Río de la Plata area (RPA) (Fig. 1). Four sampling locations were selected: Isla Santiago, in Ensenada district, Isla Paulino and Los Talas, in Berisso district, all in the south of RPA, and the Isla Martín García, located in the upper part of the Rio de la Plata. Martin García homegardens are recents. Those of Ensenada and Berisso are linked by long standing common traditions to the local context.

Present districts of Ensenada, Berisso and La Plata originated at the ends of XIXth century, as a result of the siting in 1882 of the new capital of Buenos Aires province: La Plata city. Ensenada originated as a port, being founded in 1801; on the other hand, the origin of Berisso is industrial with the settlement of leather salting houses in the decade of 1870. These last two districts were separated by the channel of access to the port of La Plata, constructed in 1883. This channel divided in two parts an ancient single island: in Ensenada district remains Isla Santiago, and Isla Paulino in Berisso district. Furthermor, in the last years of that century, an intense immigratory flow from diverse European countries —

specially from Italy, Spain, Portugal and Poland— settled in the area. Those immigrants worked in the building of La Plata city and its port and they settled in fiscal coastal lands given with the condition of developing fruticultural and horticultural tasks; there, immigrants put their own origin traditions (García, 2010; Michellod, 2000; Sanucci, 1972). So, the production of *vino de la costa* ('wine of the coast') was started, by the introduction of American grape, *Vitis labrusca* L., with vine arbor system adapted to local conditions (Marasas & Velarde, 2000). In the first half of XXth century, industrial development of RPA (meat processing plants, textile factories, petrochemistry, shipyard), as well as the continuous floodings of River of la Plata, made that many inhabitants moved to neighboring urban areas – in expansion along with industrialization- because those areas offered new job opportunities. As a consequence, they abandoned horticultural practices (García, 2010). The depopulation process was notorious in certain places, like Isla Paulino: during its splendor time, more than 70 families lived there, while only 7 families live today permanently in the island (Buet *et al.*, 2010). The initial expansion of cultivated lands caused the retraction of original vegetation composed by woods and hygrophile shrub (Cabrera, 1949). After the abandonment, native species colonized old cultivated zones, areas where also exotic species lasted or even entered and became naturalized. Changes in vegetal physiognomy, with pulses of retraction and expansion, in correlation with the rhythm of space use (cultivation, urbanization) were (and are) common in the periurban context of the RPA (Hurrell, 2008). Wine producing in particular showed an important deterioration since the second half of XXth century, in contrast with its former expansion. Nevertheless, in the last years a reactivation of wine industry is taking place, as a result of the constitution of cooperatives in Berisso district (Marasas & Velarde, 2000; Velarde *et al.*, 2008). Those initiatives recover the activity and, simultaneously, recover those traditions that are expressive of their own origin.

In Isla Martín García, like Ensenada and Berisso, the urban areas and the spontaneous vegetation suffered expanding and contracting pulses, one respect to each other. Since 1973, with the signing of the treaty of Río de la Plata between Uruguay and Argentina, this island became a natural reserve, as well as an historic site (Llambí, 1973). As it is a strategic geopolitical point, the island was under military control up to 1985, when it was transferred to the administration of Buenos Aires province as the Argentine Army retired once and for all. In its borders, in addition to scrubland, shrub and hygrophilous woods, a marginal forest similar to the *Paraná Delta* is present, with major floristic complexity than the forest found in its austral border, Punta Lara, in Ensenada district. Unlike Berisso and Ensenada, horticulture is not a typical practice of this island, where the main economical activity is tourism. Thus, homegardens are small parcels that have been settled in the last 35 years.

2. Materials and methods

This research is framed in the scope of Ethnobotany, conceived in its broadest sense as the study of the relationships between humans and their vegetal surroundings (Hurrell, 1987; Albuquerque & Hurrell, 2010). Data here presented have been obtained in successive field trips to both areas here considered, since 2001 up to date, and they are also based in observations performed in previous studies carried out in LEBA, as a part of other research projects. In the case of NOA homegardens, 44 domestic units (DU) have been visited, while in RPA, 16 DU have been studied. The total of interviewed informants reaches 44 in NOA and 25 in RPA including individuals of different gender and age, all of them involved in horticultural activity and in the elaboration of derived products, ways of consumption and eventual commercialization.

Ethnobotanical data were obtained according to usual qualitative methods (Albuquerque & Lucena, 2004; Alexíades & Sheldon, 1996; Martin, 2004) designed according to specific criteria for homegardens study (Albuquerque *et al.*, 2005; Das & Das, 2005; Vogl *et al.*, 2004). Open and semi-structured interviews to adult individuals of both gender involved in horticultural activites have been performed, as well as free listing to record the inventory of cultivated *ethnospecies* (that is, locally recognized discontinuities in plant kingdom) and varieties, life histories in the case of old people in NOA, and systematic and participant observations in different spaces in which daily life activities are developed (Pochettino, 2010). When it was possible, ethnobotanical treks were made accompanying local people during their daily activities (King, 2000; Martínez & Pochettino, 1999).

The collected information refers to ethnospecies and varieties that are present in homegardens, as well as to their reproductive material, locally recognized characteristics, production and management techniques, involved social actors, and the ascribed value and updated knowledge. In all the cases voucher specimens and other samples of involved plant species have been collected. They have been botanically identified and deposited as a document in the LEBA collections.

3. Results

3.1 NOA homegardens

The recorded ethnospecies and varieties in the homegardens of the four settlements are presented in Table 1. Local and scientific names are given, as well as the location where they have been found and the uses given by the population. The total of ethnospecies is of 41 plants, belonging to 16 botanical families. The family represented by the largest number of ethnotaxa is Cucurbitaceae: 8 (19.52%), then Solanaceae family is ranked: 6 (14.63%); Leguminosae and Chenopodiaceae are in the next place, each represented by 4 ethnotaxa (9.76%) and then Poaceae and Lamiaceae: 3 (7.32%). These families comprise 82.30% of the registered plants and they represent fundamentally the species used in food as vegetables. In the homegardens of El Shincal and Rachaite-Coranzulí the greatest diversity has been found (20 cultivated ethnospecies in each of the zones) while in Santa Victoria Oeste, only 10 ethnospecies have been recorded.

The term *homegarden* has been seldom used in NOA ethnographic studies (García *et al.*, 2002; Ottonello & Ruthsatz, 1986); and it is much more frequent the use of the local term *rastrojo* that generally designates small plots for the cultivation of food plants, while the term *potrero* is given to pasture fields (García *et al.*, 2002; Goebel, 1998; Martínez & Pochettino, 2004; Merlino & Rabey, 1978). However, the authors consider it is apropriate to give the name of *homegardens* to the small productive spaces of DU, as this category includes aspects that have not been treated in NOA studies in general, and those of *Puna* in particular, being considered an area of low biodiversity (Muscio, 1999).

In El Shincal 19 homegardens have been analyzed. Most of its inhabitants possess homegardens to self-consumption, and in several cases, *fincas* that are larger productive spaces devoted to commercial cultures, mainly walnut, but also vegetables as lettuce, condiments (cumin, anise) and pulses, for instance common bean (*Phaseolus vulgaris* L.). With regard to this species, it has to be mentioned its importance in Andean agricultural tradition as a stapple food. Though this use is still locally current, people do not plant them in homegardens, and only use those devoted to market to culture common beans, consequently generating a high homogeneity in genetic material (Lema, 2009). So, homegarden, *rastrojo* and *finca* are the categories applied by local inhabitants to name productive spaces of different magnitude, the second of them referring to cultivation areas linked to succession in the use of productive spaces. Certain DU also have *puestos* – productive spaces and second house in higher zones– because of climatic requirements of microthermal cultures and cattle pasture cycles, where homegardens are very unusual, that instead are present in the house of the valley. Each DU possesses only one homegarden, of an average size of 2m x 8m, delimited with perishable elements like shrub branches or canes. Water availability is very scarce, so the water for the cultures is provided by rain or by hand irrigation. On the contrary, the *finca* has irrigation channels, and occasionally a derivation can be made, if the homegarden is located nearby. In the homegarden food plants consumed by the members of DU (on-farm), in addition to ornamental and medicinal ones, and the criteria to plant them are exclusively familiar, for instance culinary preferences, sometimes related with communal festivities (as the case of the cultivation of white corn, a cultivar of *Zea mays* employ for the confection of *humita* – a tradicional Andean meal prepared with grated corn– during Holy Week) or family horticultural practices, as is the case of transplanting wild chile, *quitucho* (*Capsicum chacoënse* Hunz.) from the *monte* (surrounding wild vegetation) to the homegarden. This is an outstanding difference with *finca* and even with *rastrojo*, where management criteria are imposed by those who will buy the production, either other individual of the community with who exchange is made, or local market. Taking into account personal characteristics of interviewed people, differences were observed as regard of attitudes on horticulture. On the one hand, according to gender of social actors, they perform different activities: while sowing and harvest are mainly achieved by men (particularly in *finca* and *rastrojo*), women are in charge of post-harvest processing to be used in meals preparation. This fact, along with women being responsible of homegarden maintenance, results in that decisions are taken by women, who project the management of domestic residential sphere to the domestic productive one, represented by homegarden. On the other hand, an important tool was the record of life histories; this strategy allowed the authors to know, for instance, the particular devotion of one male inhabitant to experiment and register the dynamics of his cultures, much more precisely than other gardeners in El Shincal. As a result of his practices, he restored a neglected species, *poroto de manteca* (*Phaseolus lunatus* L.) and a new one, pear (*Pyrus communis* L.) in his homegarden, that was soon spread among the population because he gave seeds or cuttings to his neighbors. This example shows the crucial influence that some individuals can have in relation with the increase of diversity in the whole community, moved by their own motivations and not by the assignation of social roles. Besides, it is notorious that the space in homegardens, is the best place to carry out this experimental trials.

In Rachaite and Coranzulí, homegardens, that are called *rastrojos*, are located at higher altitude than the previous. As in El Shincal, they are exclusive for each DU and they are placed by the house. The products also are devoted to self-consumption, the informal exchange with other families, and to a lesser extent with members of nearby communities.

Each DU is the unity of ownership, production and consumption, and is constituted by a nuclear or extended family, makes its own decision on the homegarden management, without external influences. This means that there is no a communal controlling authority over these spaces, as it has been mentioned previously for other cases in Andean scope (Mayer, 2004) or elsewhere (Martínez et al., 2003). In both communities several DU also have cultivation areas for household consumption, or exchange with neighbors, in the slopes of the mountains. They are placed in ravines with natural sources of water and protected from wind. In Rachaite, cultivation areas are bigger than in Coranzulí and they are constituted by large terraces. From the study of 15 homegardens, 3 different kinds can be distinguished:

Homegardens at the same level and in the open are plots of variable size, never larger than 10 m², where plants are sowed in defined sectors. Irrigation is made by hand or by means of a pipe from a near sink. Sometimes, they are partially delimited by walls made of adobe bricks or adobe and stone (*tapia*) which is topped with mud mixed with branches of *tola (Parastrephia lepidophylla* (Wedd.) Cabrera), to avoid the erosion caused by strong winds. The rest of the plot can be delitimited and protected from animals by a wire fence. In other cases, homegardens are completely surrounded by walls that act as thermal regulators. In this kind of gardens it is frequent to fin native wild plants, besides *tola,* like *cortadera (Cortaderia* spp.), *coba (Parastrephia quadrangularis* (Meyen) Cabrera) and *añagua (Adesmia horrida* Gillies *ex* Hook. & Arn.), that are tolerated while the space they are covering is not necessary. Those native plants are sometimes used as forage, so cattle are let get into the homegarden.

The *homegardens in greenhouse* have been established since the decade of 1980, mainly because of the influence of different government organisms that irregularly contributed with seeds (generally of exotic comercial species), plastic sheets and wood necessary build up the greenhouses. In them, there is greater specific diversity than in homegardens in the open. They are *ca.* 4 m x 3 m, the walls are made of adobe and the roof of wood braces and transparent nylon. Dejections of sheeps, goats or llamas are used as fertilizer. Inside them, temperature is very high because of strong sun radiation typical of *Puna*, consequently a sort of windows are made to eliminate excessive heat in summer, or even nylon roof is replaced by metallic screen that let air the greenhouse while prevent the access of birds. Thus, greenhouse changes to a homegarden in the open and also summer association of plants changes regarding the winter one, as microenvironmental conditions are different.

The *homegardens in slopes and in the open* are constituted by small cultivation terraces delimited by a peripheral *pirca*, sometimes along with a wire fence. Each portion of terrace, where a single food plant or an association of different plants can be foun, is named *patía* and it is the unit of culture rotation. There are irrigation channels made of earth that takes the water from the nearby river. Native plants as *tola* and *cortadera* also grow naturally in the terraces, which are not eliminated unless the space they are covering is necessary. In order to protect cultures from frost, some gardeners cover a potato plant (considered the *mother of potatoes*) with a reversed vessel, because they consider that by protecting an outstanding specimen, they are protecting all the other plants of the garden.

Both in Rachaite as in Coranzulí there are homegardens at the same level and in the open, and greenhouses. Homegardens in slopes an in the open are found exclusively in Rachaite. Cultivated species are mentioned in table 1. It is noteworthy that the three kinds of homegardens here described are different from large scale production areas with commercial purpose in diverse aspects: the greater diversity of species, the presence of plants for diverse purposes (food, medicine, forage), specimens belonging to different sowing cycles, and plants with different degree of association with human beings. Though Coranzulí is a more concentrated village than Rachaite, that has a spread pattern, its inhabitants show a marked mobility among close settlements, a frequent fact in the *Puna* in particular, and in the Andes in general (Goebel, 1998; Mayer, 2004; Merlino & Rabey, 1978). This mobility is characterized by the successive occupation of the same residence by different members of extended family all along the year, or because of the temporal abandonment of the house. This dynamics has repercussions on the development of homegardens, which can be abandoned during periods when native plants or weedy forms of crops can grow. If the house is occupied by people that are not members of the same family, homegardens can be subjected to new management and criteria that though are shared in general by all the extended family, may assume particular features according who are occupying the house in the different moments of annual cycle. This is a demonstration that the structure and dynamics of homegardens are a material reflection of the structure and dynamics of the members of DU and their residence spaces.

Homegardens in Santa Victoria Oeste are also named *corralcitos* (`small corral') and they are placed very close to the house, near or adjoining to it. As in the previous cases each DU has only one homegarden. The local settlers place their houses in the bottom of the valley or in the low slopes, and they also have in the top of the slopes a minor temporal residence, called *puesto*, devoted to gardening and cattle raising where members of the DU (generally men) to supervise production. As in the case of El Shincal, both settlements are a part of the same DU. Both homegardens as the plots in *puestos* are devoted to family consumptions and local barter. The results obtained from 10 homegardens are here presented. Horticulture is developed in fluvial terraces, by profiting of the fertile soil layer which has *ca*. 40 cm thick. The garden is, generally, delimited by a dry *pirca* and has a subrectangular shape. Sometimes, inside the homegarden, small round structures of 80 cm diameter can be found, also demarcated by *pirca* of *ca*. 60 cm high. They are used as seedbeds (mostly to grow *tomate, lechuga, repollo* and *zapallito tronquero*, see Table 1), from where young plants are then moved to the central space of the garden. Irrigation is made by hand or by a pipe connected to the communal source of water. Cattle dejections are used as fertilizer and sometimes a better kind of earth is brought from elsewhere to enrich the local one. Seeds and propagules are mainly conserved from the own production. These characteristics are also mentioned for other communities in the Upper Basin of Bermejo River (Hilgert 2007a, 2007b; Hilgert & Gil 2006). In the homegarden ornamental plants (mainly flowers) are grown, besides vegetables. According to local people statements, medicinal plants are not cultivated because they do not grow well in the zone. A remarkable fact is the great diversity of varieties of corn observed in homegardens and in *puestos*, for instance *maíz blanco, maíz pisancalla*, among others, similar to the ones cultivated in El Shincal. As in this case, some varieties are exclusively grown in homegardens by the house because women prepare specific meals. Nevertheless, previous ethnobotanical studies put in evidence that, though local people sow a great diversity of crops of American origin, since 1982 there is a considerable increase of introduction of exotic plants (especially vegetables) that are replacing local cultures (Hurrell, 1990; Zardini & Pochettino, 1984). This situation was confirmed in the census conducted during this research: the amount of American cultivated species is always lower than the exotic ones.

Local names	Species	Families	Sampling sites	Local uses
Acelga	*Beta vulgaris* L. var. *cicla* L.	Chenopodiaceae	Rachaite, Coranzulí, El Shincal	Food
Achojcha	*Cyclanthera pedata* (L.) Schrader	Cucurbitaceae	Santa Victoria Oeste	Food
Ajenjo	*Artemisia absinthium* L.	Asteraceae	Rachaite, Coranzulí	Medicinal
Ají	*Capsicum annuum* L.	Solanaceae	El Shincal	Condiment
Albahaca	*Ocimum basilicum* L.	Lamiaceae	Santa Victoria Oeste,	Condiment
Alfa	*Medicago sativa* L.	Leguminosae	Rachaite, Coranzulí	Forage
Aloe vera	*Aloe vera* (L.) Burm. *f.*	Asphodelaceae	Rachaite, Coranzulí	Medicinal
Anco or Anquito	*Cucurbita pepo* L.	Cucurbitaceae	El Shincal	Food
Angola	*Cucurbita pepo* L.	Cucurbitaceae	El Shincal	Food
Arveja	*Pisum sativum* L.	Leguminosae	Rachaite, Coranzulí	Food
Cebada	*Hordeum vulgare* L.	Poaceae	Rachaite, Coranzulí	Food
Cayote	*Cucurbita ficifolia* Bouché	Cucurbitaceae	Santa Victoria Oeste	Food
Cebolla	*Allium cepa* L.	Alliaceae	Rachaite, Coranzulí, El Shincal	Food Condiment
Collareja	*Solanum tuberosum* L. subp. *andigena* (Juz. & Bukasov) Hawkes	Solanaceae	Rachaite, Coranzulí	Food
Coreanito	*Cucurbita pepo* L.	Cucurbitaceae	El Shincal	Food
Durazno	*Prunus persica* (L.) Batsch	Rosaceae	El Shincal	Food
Haba	*Vicia faba* L.	Leguminosae	Rachaite, Coranzulí	Food
Lechuga	*Lactuca sativa* L.	Asteraceae	Santa Victoria Oeste, Rachaite, Coranzulí, El Shincal	Food
Maíz	*Zea mays* L. cvs. "blanco", "blanco chico", "mediano", "diente de caballo", "pichingo", "socorro", "amarillo", "ocho rayas", "pisancalla"	Poaceae	Santa Victoria Oeste, Rachaite, Coranzulí, El Shincal	Food
Menta	*Mentha spicata* L.	Lamiaceae	Rachaite, Coranzulí	Medicinal
Morrón	*Capsicum annuum* L.	Solanaceae	El Shincal	Food

Muña-muña	*Clinopodium gilliesi* (Benth.) Kuntze	Lamiaceae	El Shincal	Medicinal
Nuez criolla	*Juglans regia* L.	Juglandaceae	El Shincal	Food
Oca	*Oxalis tuberosa* Molina	Oxalidaceae	Rachaite, Coranzulí	Food
Papa of various kinds	*Solanum tuberosum* L. subsp. *tuberosum* cvs.	Solanaceae	Rachaite, Coranzulí	Food
Papa lisa	*Ullucus tuberosus* Caldas	Basellaceae	Santa Victoria Oeste	Food
Pera	*Pyrus communis* L.	Rosaceae	El Shincal	Food
Perejil	*Petroselinum crispum* (Mill.) Fuss	Apiaceae	Rachaite, Coranzulí	Medicinal Condiment
Poroto de manteca	*Phaseolus lunatus* L.	Leguminosae	El Shincal	Food
Quínoa blanca	*Chenopodium quinoa* Willd.	Chenopodiaceae	Rachaite, Coranzulí	Food
Quínoa rosada	*Chenopodium quinoa* Willd.	Chenopodiaceae	Rachaite, Coranzulí	Food
Quitucho	*Capsicum chacoënse* Hunz.	Solanaceae	El Shincal	Condiment
Remolacha	*Beta vulgaris* L. var. *vulgaris*	Chenopodiaceae	Santa Victoria Oeste	Food
Repollo	*Brassica oleracea* L. var. *capitata* L.	Brassicaceae	Santa Victoria Oeste	Food
Ruda	*Ruta graveolens* L	Rutaceae	El Shincal	Medicinal Ornamental
Tomate	*Solanum lycopersicum* L.	Solanaceae	Santa Victoria Oeste, Rachaite, Coranzulí, El Shincal	Food
Trigo	*Triticum aestivum* L.	Poaceae	Rachaite, Coranzulí	Food
Zanahoria	*Daucus carota* L.	Apiaceae	Rachaite, Coranzulí, El Shincal	Food
Zapallito tronquero	*Cucurbita maxima* Duchesne subsp. *maxima* var. *zapallito* (Carriére) Millan	Cucurbitaceae	Santa Victoria Oeste	Food
Zapallo plomo	*Cucurbita maxima* subsp. *maxima*	Cucurbitaceae	El Shincal	Food
Zapallo silpingo	*Cucurbita maxima* subsp. *maxima* cv. "zipinka"	Cucurbitaceae	El Shincal	Food

Table 1. NOA homegardens, recorded ethnospecies, local names, botanical names, families, sampling sites and local uses

Fig. 2. A typical NOA homegarden (Santa Victoria Oeste)

3.2 RPA homegardens

In this area, homegardens are generally called *huertos* and less frequently, *quintas*. In all the cases they are located near the house of the family (that is nuclear), and usually constitute the unique place for horticulture practices, although some DU may have small places inside it. Food plants are placed in the central sector of the homegarden, and in some cases there is a special sector for aromatic and/or medicinal plants. In the homegarden a few specimens of fruit trees are also present, either disperser or in small groups. In Berisso district, the locations of Los Talas and Isla Paulino homegardens are associated to vineyards that emerged in the last years as a result of the effort of regional cooperatives. In Isla Paulino, vineyard sector have a little development, they are adjoining to homegardens forming a single unit, where is difficult to distinguish among both sectors. On the contrary, in Los Talas, vineyards occupy bigger plots and homegardens are nearby them. In all the cases men are in charge of horticultural tasks; women cooperate in that practice but their action scope is in the house. The production is destined to household consumption, but sometimes homemade products (marmelades, preserves, sauces, liquors), without additives, are prepared and then sold by direct sale (Turco *et al.*, 2006), *in situ* or in the neighboring urban areas. Vineyards are set aside for production and sale of *vino de la costa*, and at present diverse organizations are trying to introduce this product in formal marketing. The reproductive material is obtained from the own production, or *by* seeds given by the national government program Pro-Huerta from Instituto Nacional de Tecnología Agropecuaria have been used (INTA, 2011).

The ethnospecies recorded in 16 homegardens are presented in Table 2. A total of 80 species, subspecies and varieties, belonging to 27 botanical families have been recorded. The family with more taxa is Rosaceae: 11 (13.75%); it is followed by Cucurbitaceae and Lamiaceae, with 8 taxa each of them (10%), then Asteraceae, Leguminosae and Rutaceae, with 5 taxa each of them (6.25%), and Alliaceae and Solanaceae, that are represented by 4 taxa each family (5%). These families comprise 62.5% of the recorded taxa and in them, all the different kind of cultures characteristic of homegarden are represented: vegetables and pulses (Alliaceae, Asteraceae — pro parte — , Cucurbitaceae, Leguminosae, Solanaceae); medicinal and aromatic plants (Asteraceae — pro parte — , Lamiaceae, Verbenaceae) and fruits (Rosaceae, Rutaceae). In Isla Santiago homegardens the greater richness of cultivated taxa has been found: 71; while in Isla Paulino 49 taxa have been recorded, 47 taxa in Isla Martín García, and 40 in Los Talas. In this last location, where the lowest quantity of taxa has been registered, vineyards have reached the highest development. The species most frequent in all homegardens are vegetables: onion, tomato, beet, lettuce and corn. The main aromatic plants are: basil, lemon balm, mint and rosemary. The most common fruit trees are plum trees and *Citrus* tree. As regard to the production of *vino de la costa*, the cultivar of *Vitis labrusca* named "Isabella" is employed, but its fruits are not eaten.

Isla Paulino, of *ca.* 1300 ha in Berisso district, despite being only 10 km far from La Plata city, shows a precarious situation, without current services of electricity, gas and drinkable water. There is only one public telephone, but no school and only one sanitary post is open during summer. The transportation is fluvial with restricted schedules and often it is interrupted (Buet *et al.*, 2010). According to Prefectura Naval Argentina, there live 32 people in 13 houses. The 3 studied homegardens are in the open, water is obtained from the river, by means of manual irrigation, with buckets or pipes, and some rudimentary channels have been constructed. Flooding is a constant threat and when it arrives, it causes a direct damage. Homegardens average size, strictly considered, is of 25 m x 10 m. The cultivated surface is limited by both wire fences and surrounding woody vegetation. The basis of horticultural venture is local people efforts to recover a productive activity originated in the practices brought by European immigrants arrived to the zone, which allows them daily sustenance and the obtaining of some money by means of direct sale of homemade products. Vineyards are the result of cooperative work that is trying to revitalize *vino de la costa*, former largely spread. Most of food species are commonly used and are present in all homegardens. Nevertheless, there are specific cultures, as is the case of *Sechium edule* (Jacq.) Sw., what is grown in only one homegarden because of family tradition.

Los Talas, also in Berisso district, contrarily to Isla Paulino, counts on electricity, drinkable water, phone and gas, and their inhabitants — a total of 494 in 2001 (INDEC, 2011) — , access by different pathways to schools and health centers, as well as the cities of Berisso, Ensenada and La Plata (Hernández *et al.*, 2009). In this place, study area corresponds to *ca.* 300 ha, as only coastal area was considered, where homegardens and vineyards are located. As the latter are predominant, the 3 considered homegardens are smaller than those of Isla Paulino, and one of them includes a greenhouse. Water is distributed by channels, but irrigation is performed by hand. Production destiny is self-consumption.

Isla Santiago, in Ensenada district, is 800 ha large and it is 15 km far from La Plata city. In contrast to Isla Paulino, this place can be accessed either by river or by land, so a bus line from La Plata arrives there. In Isla Santiago there is initial and primary school and services of electricity, phone and drinkable water, but not gas. In 2001, there were 237 inhabitants, including Army Superior and Secondary School Río Santiago (INDEC, 2011); nowadays, according to Prefectura Naval Argentina, in this island lives 193 people in 83 houses. The current status is of recovering, after decades of abandonment of horticulture, job subocupation, and consequent depopulation. The 5 analyzed homegardens are smaller than those previously presented for the area, but they are diverse in size and shape, ranging from a square of 4 m² to a rectangle of 50 m x 25 m. Due to the initiative and personal effort of one gardener, the *kiwi, Actinidia chinensis* Planch. var. *deliciosa* (A. Chev.) A. Chev., was introduced in the island. This plant was in the past an extended commercial product, pioneer in RPA, but now remains as a relict. Other innovative gardener introduced in Isla Santiago is the *mango, Mangifera indica* L. As well, in other homegarden *Sechium edule* is planted, also linked to family tradition. *Poroto japonés, Lablab purpureus* (L.) Sweet, is here cultivated for the same reason, and it was found too in Isla Martín García.

Isla Martín García, situated in the Northern extreme of Río de la Plata, is under direct jurisdiction of Government Ministery of Buenos Aires Province. So, all plots are fiscal ones and all inhabitants are public employees in activities such as services and maintenance, or they have franchises of commercial exploitations (restaurants, camping, grocery) by paying a canon. Counting on 200 ha, it has rocky basement (one of the oldest of the country) on the contrary than the other sedimentary islands of Río de la Plata, but as well, it receives continuous contributions of sediments that modify the constitution and vegetation of the coasts. It has about 200 inhabitants (INDEC, 2011), there is no gas, but services include drinkable water and electricity (with water treatment plant and power station in the island), as well as phone, hospital, school and ship line that connects it regularly to mainland. Besides by river, it can be accessed by air, as it counts on an aerodrome with a large landing area constructed when the island was property of Argentine Army. It is easily seen that in this context, horticulture is not a typical practice in Isla Martín García, which main activity is based on tourism. The 5 studied homegardens are limited to small plots opened during the last 35 years. They have a restricted surface, not bigger than 10 m on a side, manually irrigated. The production is for self-consumption and no handmade products are confectioned to be sold. It is noteworthy that in Isla Martín García, but also in Isla Paulino and Isla Santiago, people grow *Tetrapanax papyrifera* (Hook.) K. Koch that is given the name *ambay*, because of the similitude of its leaves with those of *Cecropia pachystachya* Trécul. (= *C. adenopus* Mart. *ex* Miq.). The properties locally attributed to *T. papyrifera* are the same that the correspondent to *C. pachystachya* in Northeastern Argentine, that is its area of origin: it is used as expectorant and against cough and catarrh, as well as against asthma. In all the three locations *T. papyrifera* is cultivated in the belief that it is *C. Pachystachya*, and its cultivation is linked to a knowledge based on family traditions referred to the employ or *C. pachystachya*. Nevertheless, *T. papyrifera* also has therapeutic uses (antitussive, diuretic, febrifuge, vermifuge), although it is cultivated mainly as ornamental.

Local names	Species	Families	Sampling sites	Local uses
Acelga	*Beta vulgaris* L. var. *cicla* L.	Chenopodiaceae	Isla Martín García, Isla Santiago, Isla Paulino, Los Talas	Food
Achicoria	*Cichorium intybus* L.	Asteraceae	Isla Martín García, Isla Santiago, Isla Paulino, Los Talas	Food
Ajenjo	*Artemisia absinthium* L.	Asteraceae	Isla Santiago	Medicinal
Ají	*Capsicum annuum* L.	Solanaceae	Isla Martín García, Isla Santiago, Isla Paulino, Los Talas	Condiment
Albahaca	*Ocimum basilicum* L.	Lamiaceae	Isla Martín García, Isla Santiago, Isla Paulino, Los Talas	Condiment Medicinal
Almendra	*Prunus amygdalus* (L.) Batsch	Rosaceae	Isla Paulino	Food
Aloe	*Aloe arborescens* Mill.	Asphodelaceae	Isla Paulino	Medicinal
Aloe	*Aloe vera* (L.) Burm. *f.*	Asphodelaceae	Isla Martín García, Isla Santiago, Isla Paulino	Medicinal
Ambay	*Tetrapanax papyrifera* (Hook.) K. Koch	Araliaceae	Isla Martín García, Isla Santiago, Isla Paulino	Ornamental Medicinal
Anco	*Cucurbita moschata* (Lam.) Poir.	Cucurbitaceae	Isla Martín García, Isla Santiago, Los Talas	Food
Apio	*Apium graveolens* L.	Apiaceae	Isla Martín García, Isla Santiago, Isla Paulino, Los Talas	Food
Arveja	*Pisum sativum* L.	Leguminosae	Isla Santiago, Isla Paulino, Los Talas	Food
Banana	*Musa* x *paradisiaca* L.	Musaceae	Isla Martín García, Isla Santiago, Isla Paulino, Los Talas	Food
Batata	*Ipomoea batatas* (L.) Lam.	Convolvulaceae	Isla Martín García, Isla Santiago, Isla Paulino, Los Talas	Food
Berenjena	*Solanum melongena* L.	Solanaceae	Isla Martín García, Isla Santiago, Isla Paulino, Los Talas	Food
Caléndula	*Calendula officinalis* L.	Asteraceae	Isla Santiago	Ornamental Medicinal
Caqui	*Diospyros kaki* Thunb.	Ebenaceae	Isla Santiago, Isla Paulino	Food
Cebolla	*Allium cepa* L.	Alliaceae	Isla Martín García, Isla Santiago, Isla Paulino, Los Talas	Food Condiment

Cebolla de verdeo	*Allium fistulosum* L.	Alliaceae	Isla Martín García, Isla Santiago, Los Talas	Food Condiment
Cedrón	*Aloysia citriodora* Palau	Verbenaceae	Isla Santiago	Medicinal
Cereza	*Prunus avium* (L.) L.	Rosaceae	Isla Santiago	Food
Choclo or Maíz	*Zea mays* L.	Poaceae	Isla Martín García, Isla Santiago, Isla Paulino, Los Talas	Food
Ciboulette or Cebollín	*Allium schoenoprasum* L.	Alliaceae	Isla Martín García, Isla Santiago, Los Talas	Food Condiment
Ciruela	*Prunus domestica* L.	Rosaceae	Isla Martín García, Isla Santiago, Isla Paulino, Los Talas	Food
Damasco	*Prunus armeniaca* L.	Rosaceae	Isla Martín García	Food
Durazno	*Prunus persica* (L.) Batsch	Rosaceae	Isla Martín García	Food
Escarola	*Cichorium endivia* L.	Asteraceae	Isla Santiago, Isla Paulino	Food
Espinaca	*Spinacia oleracea* L.	Chenopodiaceae	Isla Paulino, Los Talas	Food
Eucalipto	*Eucalyptus cinerea* Benth.	Myrtaceae	Isla Santiago	Medicinal
Frutilla	*Fragaria* x *ananassa* (Weston) Duchesne	Rosaceae	Isla Santiago, Isla Paulino	Food
Guinda	*Prunus cerasus* L.	Rosaceae	Isla Martín García, Isla Santiago	Food
Granada	*Punica granatum* L.	Lythraceae	Isla Santiago, Los Talas	Food
Haba	*Vicia faba* L.	Leguminosae	Isla Santiago, Isla Paulino, Los Talas	Food
Higo	*Ficus carica* L.	Moraceae	Isla Santiago, Isla Paulino, Los Talas	Food
Kiwi	*Actinidia chinensis* Planch. var. *deliciosa* (A. Chev.) A. Chev.	Actinidiaceae	Isla Santiago	Food
Laurel	*Laurus nobilis* L.	Lauraceae	Isla Martín García, Isla Paulino, Los Talas	Condiment Medicinal
Lavanda	*Lavandula angustifolia* Mill.	Lamiaceae	Isla Santiago, Isla Paulino	Medicinal
Lechuga criolla	*Lactuca sativa* L.	Asteraceae	Isla Martín García, Isla Santiago, Isla Paulino, Los Talas	Food
Limón	*Citrus* x *limon* (L.) Osbeck	Rutaceae	Isla Martín García, Isla Santiago, Isla Paulino, Los Talas	Food Condiment
Mandarina	*Citrus reticulata* Blanco	Rutaceae	Isla Martín García, Isla Santiago, Isla Paulino, Los Talas	Food

Mango	*Mangifera indica* L.	Anacardiacae	Isla Santiago	Food
Manzana	*Malus pumila* Mill.	Rosaceae	Isla Santiago, Isla Paulino	Food
Melisa or Toronjil	*Melissa officinalis* L.	Lamiaceae	Isla Santiago	Medicinal
Melón	*Cucumis melo* L.	Cucurbitaceae	Isla Martín García, Isla Santiago, Los Talas	Food
Membrillo	*Cydonia oblonga* Mill.	Rosaceae	Isla Martín García, Isla Santiago, Isla Paulino, Los Talas	Food
Menta	*Mentha spicata* L.	Lamiaceae	Isla Martín García, Isla Santiago	Medicinal
Morrón	*Capsicum annuum* L.	Solanaceae	Isla Martín García, Isla Santiago, Isla Paulino, Los Talas	Food
Naranja	*Citrus* x *aurantium* L. (cvs.)	Rutaceae	Isla Martín García, Isla Santiago, Isla Paulino, Los Talas	Food
Nuez de pecán	*Carya illinoinensis* (Wangenh.) K. Koch	Juglandaceae	Isla Paulino	Food
Níspero	*Eriobotrya japonica* (Thunb.) Lindl.	Rosaceae	Isla Santiago, Isla Paulino, Los Talas	Food
Olivo	*Olea europea* L.	Oleaceae	Isla Paulino	Food
Orégano	*Origanum vulgare* L.	Lamiaceae	Isla Martín García, Isla Santiago, Isla Paulino	Condiment Medicinal
Palta	*Persea americana* Mill.	Lauraceae	Isla Santiago, Isla Paulino	Food
Papa	*Solanum tuberosum* L.	Solanaceae	Isla Santiago	Food
Papa del aire	*Sechium edule* (Jacq.) Sw.	Cucurbitaceae	Isla Santiago, Isla Paulino	Food
Pasto limón	*Cymbopogon citratus* (DC.) Stapf	Poaceae	Isla Santiago	Condiment Medicinal
Pepino	*Cucumis sativus* L.	Cucurbitaceae	Isla Martín García, Isla Santiago	Food
Pera	*Pyrus communis* L.	Rosaceae	Isla Santiago, Isla Paulino	Food
Perejil	*Petroselinum crispum* (Mill.) Fuss.	Apiaceae	Isla Martín García, Isla Santiago, Isla Paulino, Los Talas	Food
Poleo	*Lippia turbinata* Griseb.	Verbenaceae	Isla Santiago	Medicinal
Pomelo	*Citrus* x *aurantium* L. (cvs.)	Rutaceae	Isla Martín García, Isla Santiago, Isla Paulino, Los Talas	Food
Poroto	*Phaseolus vulgaris* L.	Leguminosae	Isla Martín García	Food
Poroto de manteca	*Phaseolus lunatus* L.	Leguminosae	Isla Martín García, Isla Santiago, Los Talas	Food

Poroto japonés	*Lablab purpureus* (L.) Sweet	Leguminosae	Isla Martín García, Isla Santiago	Food
Puerro	*Allium ampeloprasum* L.	Alliaceae	Isla Martín García, Isla Santiago, Isla Paulino, Los Talas	Food Condiment
Quinoto	*Citrus japonica* Thunb.	Rutaceae	Isla Martín García, Isla Santiago, Los Talas	Food
Rabanito	*Raphanus sativus* L.	Brassicaceae	Isla Martín García, Isla Santiago	Food
Remolacha	*Beta vulgaris* L. var. *vulgaris*	Chenopodiaceae	Isla Martín García, Isla Santiago, Isla Paulino, Los Talas	Food
Repollo	*Brassica oleracea* L. var. *capitata* L.	Brassicaceae	Isla Martín García, Isla Santiago, Isla Paulino, Los Talas	Food
Romero	*Rosmarinus officinalis* L.	Lamiaceae	Isla Martín García, Isla Santiago, Isla Paulino, Los Talas	Condiment Medicinal
Rúcula	*Eruca vesicaria* (L.) Cav.	Brassicaceae	Isla Martín García, Isla Santiago, Isla Paulino	Food Condiment
Salvia	*Salvia officinalis* L.	Lamiaceae	Isla Santiago	Condiment
Sandía	*Citrullus lanatus* (Thunb.) Matsum. & Nakai	Cucurbitaceae	Isla Martín García, Isla Santiago, Los Talas	Food
Tomate	*Solanum lycopersicum* L.	Solanaceae	Isla Martín García, Isla Santiago, Isla Paulino, Los Talas	Food
Tomillo	*Thymus vulgaris* L.	Lamiaceae	Isla Santiago	Condiment
Uva americana	*Vitis labrusca* L.	Vitaceae	Isla Santiago, Isla Paulino, Los Talas	Wine
Uva europea	*Vitis vinifera* L.	Vitaceae	Isla Santiago	Food
Zanahoria	*Daucus carota* L.	Apiaceae	Isla Martín García, Isla Santiago, Isla Paulino, Los Talas	Food
Zapallo	*Cucurbita maxima* Duchesne subsp. *maxima*	Cucurbitaceae	Isla Martín García, Isla Santiago, Isla Paulino	Food
Zapallo hongo	*Cucurbita maxima* Duchesne subsp. *maxima* (cv.)	Cucurbitaceae	Isla Martín García, Isla Santiago, Isla Paulino	Food
Zapallito	*C. maxima* subsp. *maxima* var. *zapallito* (Carriére) Millan	Cucurbitaceae	Isla Martín García, Isla Santiago, Los Talas	Food

Table 2. RPA homegardens, recorded ethnospecies, local names, botanical names, families, sampling sites and local uses

Fig. 3. A typical RPA homegarden (Isla Paulino)

4. Discussion

By comparison with large scale agriculture areas, that are in general large extensions with one or a few species (homogeneous), the studied homegardens both in NOA as in RPA, are reduced spaces with a high biodiversity (highly heterogeneous), as they are spaces of multiple production, where plants with diverse applications (food, condiment, medicine, ornamental, forage) are grown.

Homegardens have differential features referred to their *materiality* and *spatiality*. Both homegardens of NOA and RPA are small plots close to the residential structures of domestic units. In NOA, homegarden is constituted by only one space, while in RPA, though the single space is the general tendency, some DU can have a parted homegarden between various plots. In RPA there are also spaces assigned to the cultivation of low number of fruit trees. In some of the studied sited of this last area, homegardens are linked with vineyards, that occasionally may have not a high development and are adjoining to homegardens, so they are considered as only one unit. Other times, vineyards occupy larger extensions and homegardens are then adjoined to them. In NOA, homegardens are never nearby the larger areas of commercial cultivation, which are also separated from DU residential space, to take advantage of the different microenvironments that are found in the area.

In homegardens, *selection criteria* and *management practices* are decided by the own DU, by contrast with the areas of commercial cultivation that are governed by market, other DU or other communities members. In the case of NOA, homegardens production is exclusively employed for self-consumption, only in a few exceptions the members of DU exchange vegetables, seeds or propagules, mostly after the successful introduction of new varieties. In RPA homegardens, although the major part of the production goes to household consumption, occasionally homemade products are marketed at a low scale by means of direct sale. In RPA the origin of current cultures is the own production of DU, or informal exchanges, even the contribution of agencies not linked to local communities, for instance government organisms, but there are also several cases where the proceeding of determined species can be only explained by family traditions: *Sechium edule* (Isla Santiago, Isla Paulino),

Lablab purpureus (Isla Martín García, Isla Santiago). This results on a great agrodiversity, as a consequence of different traditions in vogue conforming the pluricultural context of the area. A similar situation is observed in NOA, where the presence of certain species is linked to family wisdom, either culinary (for instance, dishes prepared with particular varieties of corn in El Shincal and Santa Victoria Oeste; some of them in occasion of festivities as is the case of *humitas* prepared during Holy Week in El Shincal) or therapeutic ones. In adittion, in NOA people conserve productive technology and architecture (*patías*, in Rachaite; homegardens with adobe walls as thermal regulators in Coranzulí) which history can be traced in the prehispanic past of the region. This Andean tradition is also expressed in a high infraspecific diversity of American crops anciently domesticated. It is necessary to think about personal motivations in these situations of change, that, as it was previously exposed, have led to the restoring and revalorization of relictual crops in the homegardens.

Homegardens are also *dynamic spaces*. In both study areas, it is observed that they can change their physiognomy (covered or not), and as a consequence their microclimatic conditions, as well as the association of species, they may be as well abandoned during certain time to be then used again, and it is usual to find specimens planted in different moments of horticultural cycle, all of them in different growth status. The dynamics of this productive area is assembled with the UD one, as when members move from a residence to another, selection criteria also change. Therefore, it is an artificial space, a social artifact (Mayer, 2004) where the relationship human beings-plant is flexible, subject to local contingencies and criteria, shaped by each DU peculiarities, which contributes to its differential perdurability.

Homegardens are *spaces of innovation and experimentation* for the cultivation of new species or varieties (for instance pear in NOA, kiwi and mango in RPA). This aspect also has been recorded in both Peruvian and Chilean Puna (Aldunante *et al.*, 1981; Mayer, 2004), where general characterization of homegardens is similar to the one given here (Harris, 1969, 1989). Furthermore, it has to be mentioned the incorporation of new technologies and seeds given by organisms foreign to local community, what is noticeable in the physical spaces occupied by homegardens in both study areas. Both in Coranzulí and Santa Victoria Oeste in NOA, as in RPA, by means of the program Pro-Huerta developed by the national agency for the development of agriculture (INTA) seeds, other consumables, and technical advice were given to gardeners. This program is enclosed in the national plan for food security, which is supposed to guarantee a diversified nourishing, by means of the self production of fresh food (INTA, 2011). The newly incorporated elements are then adapted to local characteristics, particularly in NOA where the weather does not fit in the general patterns employed in national programs. This fact is easily seen in structural changes made in greenhouses in Coranzulí (nylon roofs were replaced by metallic screen) and the seedbeds in Santa Victoria Oeste, where this new sowing practice has been technologically adapted to local productive architecture.

Homegardens can be considered *reservoirs of plant varieties with different degree of association with human beings*, what is evidenced in the simultaneous presence in homegardens of cultivated specimens, either domesticated or not, associated with weedy and wild plants. So, homegardens can also be seen as *reservoirs of cultural practices of management*. NOA homegardens (in particular those in the high) constitute a clear example of productive spaces where plant species are handled: cuttings, sowing, care, tolerance and eradication of different plants are done, without producing a uniform population of domesticated plants. On the contrary there exists a wide range of botanical elements which relationship with human beings is closer or looser according to reproductive autonomy and phenotypic associated changes. The authors assume that this situation emerges from the very close relationship held by gardeners with the plants that they select. This relationship is characterized by a low degree of environmental disturbance, and the management of a diverse set of plants that are cultivated, tolerated and supported against market tendencies, and, in very few cases, they are refused and or eradicated. That is why in a homegarden there are many cultivated plants, several wild ones and very few weeds.

5. Conclusions

Horticulture carried out in homegardens, by means of the study of its practices, is a valuable tool to evaluate the BK that guides them. Homegardes constitute a propitious sphere to develop ethnobotanical studies, because of their reduced scale and the proximity of the productive plot to the residential segment, both of domestic unit. Nevertheless, they show a great diversity in species and varieties, which destiny is not only as food, but condiment, medicine and ornament as well. This diversity is expressive of the different degrees of relationship of family group with their plant resources: when considering weeds that accompany some crops, this diversity increases. Therefore, homegardens are spaces where BK becomes *visible* (TBK or BK linked to traditions), where knowledge and beliefs are embodied in strategies and practices, through the direct contact between human beings and their environment, what shapes a dynamic and versatile relationship between people and plants (Pochettino & Lema, 2008).

On the one hand, in areas with an ancient tradition of occupation and horticultural activity, as the case of rural communities in NOA reflects, the relationship is explained by homegarden domestic character, as it is the productive aspect of DU (while house is the residential aspect), that in turn prints its own features in homegardens: its members history, the dynamics of occupation of the house, selection criteria of species and management practices of cultivation plots linked to local/familiar traditions as well as personal preferences. For these reasons, homegardens constitute a particularly rich field to ethnobotanical studies devoted to local BK from the perspective of the own involved actors.

On the other hand, periurban sectors of RPA are part —for a combination of geographic as historical, economical and social reasons— of the complex conurbation Buenos Aires-La Plata, the biggest one in Argentina, both in extension as in population. In this frame, this work contributes to the development of Urban Ethnobotany, as it analyzes, within the characterization of *urban botanical knowledge* (Hurrell *et al.*, 2011; Pochettino *et al.*, 2011), the components linked to traditions of different origin (groups of immigrants, familiar preferences, culinary wisdom and therapeutic practices), which remain invisible for most of the urban population, but are the basis of horticultural activity in RPA. Thus, horticultural practices in periurban homegardens rely largely on family traditions, so they are guided by a BK linked to traditions, but not traditional itself. Because of their own functionality, homegardens are dynamic systems because the BK that guides their practices becomes from dynamics adaptations to specific conditions. In NOA horticulture, in a homogeneous cultural context, BK is mainly *traditional*. In periurban horticulture, in a major pluricultural context of the conurbation where it is immersed, the current BK is *linked to traditions*. In both cases, BK is *adaptative* as it allows the gardeners to make adjustments to environment changing conditions, by means of the orientation of their strategies of selection and use of plants. Consequently, their study and preservation are necessary.

6. Acknowledgments

The authors acknowledge to Carina Llano for her assistance with figures and to Emilio A. Ulibarri (Instituto de Botánica Darwinion, CONICET-Academia Nacional de Ciencias Exactas, Físicas y Naturales), Fernando Buet Costantino and Jeremías P. Puentes (LEBA) for their cooperation during field work. Local inhabitants of all communities are specially acknowledged for their warm hospitality and generosity in sharing their time and wisdom, as well as their consent to publish the results of the research. Otherwise, this work would not be achieved. The research was conducted with the financial support of CONICET and UNLP.

7. References

Albeck, M. E. (2003/2005). Sitios agrícolas prehispánicos: la búsqueda de indicadores cronológicos y culturales. *Cuad. Inst. Nac. Antropol. y Pensamiento Latinoamericano* Vol. 20, pp. 13-26, ISSN 0570-8346.

Albeck, M. E., Lupo, L., Maloberti, M., Pigoni, M., Zapatiel, J., Korstanje, A. & Cuenya, P. (2008). An interdisciplinary approach for Coctaca: Stimulating results for the comprehension of an ancient agricultural complex. *VII Int. Meeting Phytolith Res. IV Southamer. Meeting Phytolith Res.*, pp. 59. Mar del Plata.

Albuquerque, U. P. de & Hurrell, J. A. (2010). Ethnobotany: one concept and many interpretations. In: *Recent developments and case studies in Ethnobotany*, Albuquerque, U. P. de & Hanazaki, N. (eds.), pp. 87-99. Brazilian Soc Ethnobiol & Ethnoecol/Publ Group of Ecol & Appl Ethnobot, Recife, ISBN 978-85-7716-711-1.

Albuquerque, U. P. de & Lucena, R. F. (2004). Métodos e técnicas na pesquisa etnobotánica. Livro Rápido/NUPEEA, Recife, ISBN 85-89501-26-4.

Albuquerque, U. P. de, Andrade, L. & Caballero, J. (2005). Structure and floristics of homegardens in Northeastern Brazil. *J. Arid Environments* Vol. 62, pp. 491-506, ISSN 0140-1963.

Aldunante, C.; Armesto, J.; Castro V & Villagrán, C. (1981). Estudio etnobotánico en una comunidad precordillerana de Antofagasta: Toconce. *Bol. Museo Hist. Nat. Chile*, Vol. 38, pp. 183-223, ISSN 0027-3910.

Alexíades, M. N. & Sheldon, J. W. (1996). *Selected guidelines for ethnobotanical research: a field manual*. The New York Botanical Garden, New York, ISBN 0893274046.

Balick, M. J. & Cox, P. (1996). *Plants, people and culture. The science of Ethnobotany*. Sci. Amer. Library, New York, ISBN 0-7167-5061-9.

Barsky, A. (2010). La agricultura de "cercanías" a la ciudad y los ciclos del territorio periurbano. Reflexiones sobre el caso de la Región Metropolitana de Buenos Aires. In: *Agricultura periurbana en Argentina y globalización. Escenarios, recorridos y problemas*, Svetlitza de Nemirovsky, A. (ed.), pp. 15-29. FLACSO, Buenos Aires, ISSN 2218-5682.

Blanckaert, I., Swennen, R., Paredes, M., Rosas, R. & Lira Saade, R. (2004). Floristic composition, plant uses and management practices in homegardens of San Rafael Coxcatlán, Valley of Tehuacán, Mexico. *J. Arid Environments* Vol. 57, pp. 179-202, ISSN 0140-1963.

Berberián, E. & Nielsen, A. (1988). Sistemas de asentamiento prehispánicos en la etapa formativa del Valle de Tafi. In: *Sistemas de asentamiento prehispánicos en el Valle de Tafí*, Berberián, E. & Nielsen, A. (eds.), pp. 21-55. Comechingonia, Córdoba, ISSN 0326-7911.

Boixadós, R. (2002). Los pueblos de indios de La Rioja colonial. Tierra, trabajo y tributo en el siglo XVII. In: *Los pueblos de indios del Tucumán colonial. Pervivencia y desestructuración*. Farberman J. & Gil Montero, R. (eds.), pp. 15-17, Ed. UNQ, Buenos Aires, ISBN 978-987-9173-64-0.

Buet Costantino, F., Ulibarri, E. A. & Hurrell, J. A. (2010). Las huertas familiares en la isla Paulino (Buenos Aires, Argentina). In: *Tradiciones y Transformaciones en Etnobotánica*, Pochettino, M. L., Ladio A. H. & Arenas, P. M. (eds.), pp. 479-484. CYTED-RISAPRET, San Salvador de Jujuy, ISBN 978-84-96023-95-6.

Cabrera, A. L. (1949). Las comunidades vegetales de los alrededores de La Plata (Provincia de Buenos Aires, República Argentina). *Lilloa* Vol. 20, pp. 296-376, ISSN 0075-9481.

Cabrera, A. L. (1976). Regiones fitogeográficas argentinas. *Encicl. Argent. Agric. Jard.* Vol 2, pp. 1-85. Acme, Buenos Aires, ISBN 950-566-314-5.

Capparelli, A., Lema V. y & Giovannetti, M. (2004) El poder de las plantas. In: *El Shincal de Quimivil*, Raffino, R. (ed.), pp: 140-163. Editorial Sarquís, Catamarca, ISBN: 987-9170-33-4

Capparelli, A., Lema, V., Giovannetti, M. & Raffino, R. (2005). Introduction of European crops (wheat, barley and peach) in Andean Argentina during the 16th century: archaeobotanical and ethnohistorical evidence. *Vegetation History and Archaeobotany* Vol. 14, pp. 472-484, ISSN 0939-6314.

Capparelli, A., Hilgert, N., Ladio, A. H., Lema, V., Llano C., Molares, S., Pochettino, M. L., & Stampella. P. (2011). Paisajes culturales de Argentina: pasado y presente desde la perspectiva etnobotánica y arqueobotánica *Revista En Línea de la Asociación Argentina de Ecología del Paisaje* Vol. 2, No. 2 (in press), ISSN 1853-8045

Castellano, M. (2000). Updating aboriginal traditions of knowledge. In: *Indigenous knowledges in global contexts. Multiple readings of our world*, Sefa Dei, G. J., Hall, B. L. & Rosenberg, D. G. (eds.), pp. 21-36. University of Toronto Press, Toronto, ISBN 13: 978-0802080592.

Das, T. & Das. A. K. (2005). Inventorying plant biodiversity in homegardens. A case study in Barak Valley, Assam, North East India. *Current Sci.* Vol. 89, No. 1, pp. 155-163, ISSN: 0011-3891.

Del Río, J., Maidana, J., Molteni, A., Pérez, M., Pochettino, M. L., Souilla, L., Tito, G. & Turco, E. (2007). El rol de las "quintas" familiares del Parque Pereyra Iraola (Buenos Aires, Argentina) en la conservación de la agrobiodiversidad. *Kurtziana* Vol. 33, No. 1, pp. 217-226, ISSN 1852-5962.

Gadgil, M., Berkes, F. & Folke, C. (1993). Indigenous knowledge for biodiversity conservation. *AMBIO* Vol. 22, No. 2-3, pp. 151-156, ISSN 0044-7447.

García, M. (2010). Inicios, consolidación y diferenciación de la horticultura platense. In: *Agricultura periurbana en Argentina y globalización. Escenarios, recorridos y problemas*, Svetlitza de Nemirovsky, A. (ed.), pp. 69-85. FLACSO, Buenos Aires, ISSN 2218-5682.

García, S., Rolandi, D., López, M. & Valeri, P. (2002), "Alfa", vega y hortaliza. Riego y siembra en Antofagasta de la Sierra, puna catamarqueña. *Relaciones Soc. Argent. Antropol.* Vol. 27, pp. 79-100, ISSN 0325-2221.

Goebel, B. (1998). "Salir de viaje" Producción pastoril e intercambio económico en el noroeste argentino. *Estudios Americanistas de Bonn*, Vol. 30, pp. 867-891, ISSN 0176-6546.

Harris, D. (1969). Agricultural systems, ecosystems and the origins of agriculture. In: *The domestication and exploitation of plants and animals*, Ucko, P. & Dimbleby, G. (eds.), pp. 3-15 , Duckworth, London, ISBN 9780202361697.

Harris, D. (1989). An evolutionary continuum of people-plant interaction. In: *Foraging and Farming. The evolution of plan exploitation*, D. Harris, D. & Hillman, G. (eds.), pp. 11-26. Unwin Hyman, London, ISBN 978-0044452355..

Hernández, M. P., Colares, S. M. & Civitella, S. M. (2009). Plantas utilizadas en medicina popular en un sector del Partido de Berisso, Buenos Aires, Argentina. *Bol. Latinoamer. Caribe Pl. Medic. Aromát.* Vol. 8, No. 5, pp. 435-444, ISSN 0717-7917.

Hilgert, N. I. (2007a). Plantas silvestres, ámbito doméstico y subsistencia. In: *Finca San Andrés. Un espacio de cambios ambientales y sociales en el Alto Bermejo*, Brown, A. D., García Moritán, M., Ventura, B. N., Hilgert, N. I. & Malizia, L. R. (eds.), pp. 167-228, Ediciones del Subtrópico, Tucumán, ISBN 978-987-23533-1-5

Hilgert, N. I. (2007b). La vinculación del hombre actual con los recursos naturales y el uso de la tierra. In: *Finca San Andrés. Un espacio de cambios ambientales y sociales en el Alto Bermejo.* , Brown, A. D., García Moritán, M., Ventura, B. N., Hilgert, N. I. & Malizia, L. R. (eds.), pp. 159-186. Ediciones del Subtrópico, Tucumán, ISBN 978-987-23533-1-5

Hilgert, N. I. & Gil, G. E. (2006). Plants of the Las Yungas Biosphere Reserve, Northwest of Argentina, used in health care. *Biodiversity & Conservation* Vol. 15, pp. 2565-2594.

Hilgert, N. I. & Gil, G. E. (2008). Los cambios de uso del ambiente y la medicina herbolaria. Estudio de caso en Yungas argentinas. *Bol. Latinoam. Caribe Pl. Medic. Aromat.* Vol. 7, No. 3, pp. 130-140, ISSN 0960-3115.

Hurrell, J. A. (1987). Las posibilidades de la etnobotánica y un nuevo enfoque a partir de la ecología y su propuesta cibernética. *Revista Española Antropol. Amer. (Madrid)* Vol. 17, pp. 235-258, ISSN 0556-6533.

Hurrell, J. A. (1990). *Interpretación de relaciones en ecología, a partir de la noción de sistema, en Santa Victoria e Iruya (Salta, Argentina).* Tesis Doctoral 548 (inédita). Facultad de Ciencias Naturales y Museo, Universidad Nacional de La Plata.

Hurrell, J. A. (1991). Etnomedicina: enfermedad y adaptación en Iruya y Santa Victoria (Salta, Argentina). *Revista Museo de La Plata (n.s.) Antropol.* 9 (69): 109-124. ISSN 0376-2149.

Hurrell, J. A. (ed.). 2008. *Flora Rioplatense* Vol. 3, No. 1, pp. 1-334. Ed. Lola, Buenos Aires, ISBN 978-987-1533-02-2.

Hurrell, J. A. & de la Sota, E. R. (1996). Etnobotánica de las Pteridófitas de los pastizales de altura de Santa Victoria (Salta, Argentina). *Revista Museo de La Plata (n.s.) Bot.* 14 (104-105): 151-164. ISSN 0376-2149.

Hurrell, J. A., Ulibarri, E. A., Puentes, J. P., Buet Costantino, F., Arenas, P. M. & Pochettino, M. L. (2011). Leguminosas medicinales y alimenticias utilizadas en la conurbación Buenos Aires-La Plata, Argentina. *Bol. Latinoam. Caribe Pt. Med. Aromat.* Vol. 10. No. 5, pp. 443-455, ISSN 0960-3115.

INDEC (2011). Instituto Nacional de Estadística y Censos. Argentina. VI-2011, Available from: < http://www.indec.mecon.ar>

INTA (2011). Instituto Nacional de Tecnología Agropecuaria. Argentina. VII-2011. Available from: <http://www.inta.gov.ar>

King, A. (2000). A brief review of participatory tools and techniques for the conservation and use of plant genetic resources. In: *Participatory approaches to the conservation and use of plant genetic resources*, Friis-Hansen, E. & Sthapit, B. (eds.), pp. 27-43. Int. Pl. Gen. Res. Inst., Rome, ISBN 9789290434443.

Lamont, S., Hardy Eshbaugh, W. & Greenberg, A. (1999). Species composition, diversity, and use of homegardens among three Amazonian villages. *Econ. Bot.* Vol. 53, No. 3, pp. 312-326, ISSN 0013-0001.

Lema, V. (2006). Huertos de altura: el manejo humano de especies vegetales en la puna argentina. *Revista de la Escuela de Antropología* Vol. 12, pp. 173-186. Universidad Nacional de Rosario, ISSN 1852-1576.

Lema, V. (2009). Domesticación vegetal y grados de dependencia ser humano-planta en el desarrollo cultural prehispánico del noroeste argentino Tesis Doctoral 1020 (inédita), Facultad de Ciencias Naturales y Museo, Universidad Nacional de La Plata, La Plata.

Llambí, A. (1973). *La Isla Martín García.* Ministerio de Bienestar Social, Provincia de Buenos Aires, La Plata.

Maffi, L (2001). Introduction. On the interdependence of biological and cultural diversity. In: *On biocultural diversity*, Maffi, L. (ed.), pp. 1-50. Smithsonian, Washington, ISBN 1-56098-905-X.

Maidana, J., Pérez, M., Tito, G. & Turco, E. (2005). Ecohorticultura en el Parque Pereyra, La Plata-Berazategui (Buenos Aires, Argentina). LEISA, *Revista de Agroecología* Vol. 20, No. 4, pp. 42-44, ISSN 1729-7419.

Marasas, M. & Velarde, I. (2000). Rescate del saber tradicional como estrategia de desarrollo: los viñateros de la costa. *Bol. ILEIA* Vol. 16, No. 2, pp. 23-24.

Martin, G. J. (2004). *Ethnobotany. A methods manual.* Earthscan, London, ISBN 1844070840.

Martínez, M. R. & Pochettino, M. L. (1999). El valor del conocimiento etnobotánico local: aporte a la currícula educativa en el área de biología en las escuelas de Molinos. Valles Calchaquíes, Provincia de Salta. *Cuad. INAPL* Vol. 18, pp. 257-270, ISSN 0570-8346.

Martínez, M. R. & Pochettino, M. L. (2004). Microambientes y recursos vegetales terapéuticos. Conocimiento local en Molinos, Salta, Argentina. *Zonas Áridas* Vol. 8, pp. 18-31, ISSN 1013-445X.

Martínez, M. R., Pochettino, M. L. & Arenas, P. M. (2003). La horticultura: estrategia de subsistencia en contextos pluriculturales, Valle del Cuñapirú, Misiones, Argentina. *Delpinoa* Vol. 45, pp. 89-98, ISSN 0416-928X.

Mayer, E. (2004). *Casa, chacra y dinero. Economías domésticas y ecología en los Andes*. Inst. Estud. Peruanos, Lima, ISBN 9789972511165.

Merlino, R. & Rabey, M. (1978). El ciclo agrario-ritual en la puna argentina. *Relaciones Soc. Argent. Antropol.* Vol. 12, pp. 47-70, ISSN 0325-2221.

Michellod, O. E. (2000). *La identidad del paisaje urbano a través de la memoria. Berisso, Argentina*. Ed. Al Margen, La Plata, ISBN 978-84-96023-95-6.

Muscio, H. (1999). Colonización humana del NOA y variación en el consumo de los recursos: la ecología de los cazadores recolectores de la Puna en la transición Pleistoceno-Holoceno. *Naya*, IV-2008. Available from: <http://www.antropologia.com.ar>

Nazarea, V. (1998). *Cultural Memory and Biodiversity*. University of Arizona Press, Tucson, ISBN 9780816516810.

Nielsen, A. (1996). Demografía y cambio social en la Quebrada de Humahuaca (Jujuy, Argentina) *ca.* 700-1535 d.C. *Relaciones Soc. Argent. Antropol.* Vol. 21, pp. 307-385, ISSN 0325-2221.

Ottonello, M. & Ruthsatz, B. (1986). Agricultura prehispánica y la comunidad hoy en la quebrada de Rachaite. Provincia de Jujuy, Argentina. *Runa* Vol. 16, pp. 1-27, ISSN 0325-1217.

Pérez, M., Tito, G. & Turco, E. (2008). La producción sin agrotóxicos del Parque Pereyra Iraola: ¿un sistema agroalimentario localizado en el periurbano? In: *Sistemas agroalimentarios localizados en Argentina*, Velarde, I., Maggio, A. & Otero, J. (eds.), pp. 102-121. Instituto Nacional de Tecnología Agropecuaria, Buenos Aires, ISBN 978-950-34-0493-5.

Pochettino, M. L. (2010). Huertos periurbanos como aporte a la diversidad agrícola, Provincia de Buenos Aires, Argentina. In: *Tradiciones y Transformaciones en Etnobotánica*, Pochettino, M. L., Ladio A. H. & Arenas, P. M. (eds.), pp. 186-192. CYTED-RISAPRET, San Salvador de Jujuy, ISBN 978-84-96023-95-6.

Pochettino, M. L. & Lema, V. (2008). La variable tiempo en la caracterización del conocimiento botánico tradicional. *Darwiniana* vol. 46, No. 2, pp. 227-239, ISSN 0011-6793.

Pochettino, M. L., Souilla, L. & Turco, E. (2006). Adaptación a nuevas condiciones sociales y económicas entre los viejos "quinteros" del Parque Pereyra Iraola (Buenos Aires). *VIII Congr. Antropol. Social, Simposio*, Salta (CD).

Pochettino, M. L., Puentes, J. P., Buet Costantino, F., Arenas, P. M., Ulibarri, E. A. & Hurrell, J. A. (2011). Functional Foods and Nutraceuticals in a Market of Bolivian Inmigrants in Buenos Aires (Argentina). *Evidence-Based Complementary and Alternative Medicine*, Vol. 2012 (in press), ISSN 1741-427X.

Quesada, M. (2006). El diseño de las redes de riego y las escalas sociales de la producción agrícola en el 1er. milenio DC (Tebenquiche Chico, Puna de Atacama). *Estudios atacameños* Vol. 31, pp. 31-46, ISSN 0716-0925.

Rodríguez, M. F. & Aschero, C. (2007). Confección de cordeles en la Puna septentrional y meridional argentina. Movilidad e interacciones socioeconómicas. In: *Paleoetnobotánica del Cono Sur: estudios de caso y propuestas metodológicas*, Marconetto, B., Babot, P. & Oliszewsky, N. (eds.), pp. 11-24. Museo de Antropología, Universidad Nacional de Córdoba, ISBN 978-987-1110-60-5.

Sanucci, L. (1972). Berisso, un reflejo de la evolución Argentina. Municipalidad Berisso, Berisso.

Tarragó, M. (1980). El proceso de agriculturización del Noroeste argentino, zona Valliserrana y sus relaciones con zonas vecinas. *Actas V Congr. Nac. Arqueol. Argent.*, pp. 181-217. Facultad de Filosofía, Humanidades y Artes, Universidad Nacional de San Juan.

Tarragó, M. N. (ed.) (2000). *Los pueblos originarios y la conquista. Nueva Historia Argentina*. Vol. 1, Ed. Sudamericana, Buenos Aires, ISBN 9500717921

Turco, E., Souilla, L. & Pochettino, M. L. (2006). Relación entre saberes y prácticas culinarios con la conservación de la agrobiodiversidad. Estudio etnobotánico en el Parque Pereyra Iraola (Buenos Aires, Argentina). *VII Congr. Latinoamer. Sociol. Rural*. 20-24 de noviembre, Quito (CD).

Velarde, I., Voget, C., Avila, G., Loviso, C., Orosco, E., Sepúlveda, C. & Artaza, S. (2008). Influencia de la calidad en el consumo de productos patrimoniales: el caso del sistema agroalimentario del vino de la costa de Berisso. In: *Sistemas agroalimentarios localizados en Argentina*, Velarde, I., Maggio, A. & Otero, J. (eds.), pp. 31-66. Instituto Nacional de Tecnología Agropecuaria, Buenos Aires, ISBN 978-950-34-0493-5..

Vogl, C. R., Vogl-Lukasser, B. & Caballero, J. (2002). Homegardens of Maya migrants in the Palenque District (Chiapas/Mexico). Implications for sustainable rural development. In: *Ethnobiology and Biocultural Diversity*, Stepp, J. R., Wyndham, F. S. & Zarger, R. K. (eds.), pp. 631-647. University of Georgia Press, Athens, ISBN 0-8203-2349-7

Vogl, C. R., Vogl-Lukasser, B. & Puri, R. (2004). Tools and Methods for data collection in ethnobotanical studies of homegardens. *Field Methodes* Vol. 16, No. 3, pp. 285-306, ISSN1525-822X.

Vogl-Lukasser, B., Vogl, C. R. & Bolhàr-Nordenkampf, H. (2002). Homegarden composition on small peasant farms in the Alpine regions of Eastern Tyrol (Austria) and their role in sustainable rural development. In: *Ethnobiology and Biocultural Diversity*, Stepp, J. R., Wyndham, F. S. & Zarger, R. K. (eds.), pp. 648-658. University of Georgia Press, Athens, ISBN 0-8203-2349-7.

Wagner, G. (2002). Why plants have meanings? In: *Ethnobiology and Biocultural Diversity*, Stepp, J. R., Wyndham, F. S. & Zarger, R. K. (eds.), pp. 659-667. University of Georgia Press, Athens, ISBN 0-8203-2349-7

Watson, J. W. & Eyzaguirre, P. B. (eds.). (2002). *Home gardens and in situ conservation of plant genetic resources in farming systems*. International Plant Genetic Resources Institute, Rome, ISBN 978-9290435174.

Yacobaccio, H. (2007). Complejidad social, especialización y domesticación de camélidos en cazadores recolectores Surandinos. In: *Sociedades precolombinas surandinas: temporalidad, interacción y dinámica cultural del NOA en el ámbito de los Andes centro-sur*, Williams, V., Ventura. B., Callegari. A. & Yacobaccio, H. (eds.) pp. 203-220. Buschi, Buenos Aires, ISBN 9789870532828.

Zardini, E. M. & Pochettino, M. L. (1984). Resultados de un viaje etnobotánico al norte de Salta (Argentina) *IDIA*, Vol. sep.-dic. 1983, pp. 108-121, ISSN 0018-9081.

Part 6

New Technologies

8

Postharvest Technologies of Fresh Horticulture Produce

Alejandro Isabel Luna-Maldonado[1], Clement Vigneault[2] and Kei Nakaji[3]
[1]Department of Agricultural and Food Engineering, Faculty of Agriculture,
Autonomous University of Nuevo Leon, Escobedo, N.L.,
[2]Department of Bioresource Engineering, Macdonald Campus,
McGill University, Ste-Anne-de-Bellevue, Québec,
[3]Kyushu University, Laboratory of Agricultural Ecology,
Division of Agricultural Ecology, Department of Plant Resources, Fukuoka,
[1]Mexico
[2]Canada
[3]Japan

1. Introduction

The actual processing of fresh horticultural produce and the reducing labor force have begun to force the development of robots that are capable of dealing with the variations inherent in the produce being handled, stored and transported. All around the world, there is increasing interest in the use of robots to replace the drastic decreasing number of farmers due to the lower birth rate and the increasing average age of the remaining farmers (Kitamura *et al., 2004* & Wang, 2009). In the USA, the supply of workers available for hand harvesting is decreasing steadily and true shortages are occurring. Without enough workers when needed for a few weeks each year, a large amount of hand harvested crops will be lost. Fruit and vegetable crops need new productive harvesting technologies (Sarig *et al.,* 2010)

In general, with horticulture applications, the variations of fresh produce in size and color as well as in the internal structure and defects such as over-ripeness and bruising damage are considered the main barriers to the extension of robotics. Robots have to cope with produce delivered with random orientation and placement, and with environmental issues such as a wide range of humidity and temperature resulting in condensation problems on electronic circuit or dew on sensors of all kinds. It requires manipulator mechanisms, controls and end-manipulators to be designed to interact with the surrounding environment, sensing techniques, mobility, and work cell development. Typical applications where imaging and electro-sensing processes can be used to guide robots include sensing general fruit quality (ElMasry *et al.* 2009; Li *et al.*, 2006 & Lino *et al.*, 2008), measuring internal quality parameters (Chen, 2008 & Li *et al.*, 2009), or detecting mechanical (ElMasry *et al.*, 2008) or physiological (ElMasry *et al.*, 2009) damage, for either grading the produce accordingly or involving automatic trimming robots to remove the defect. In other words, processed items are picked from a conveyor belt and presented to a cutting or trimming device to further process the produce in order to maintain a consistency of quality in the finished produce. For example, during the inspection of lettuce heads and their core removal (Key Technology, 2010), a fresh lettuce will be scanned to detect stones and insect damage. At the same time, a two directional measurement device can take images to provide core position and orientation information. Based on these pieces of information, the lettuce will be properly aligned with a blade mounted on a robot to perform the core removal (Applied Sorting Technologies, 2010).

The above example is considered as a typical robot application in horticulture. There is also a totally different type of robot application that is often forgotten due to its relative simplicity which is to fill produce in a standard package unit such as reusable plastic container (RPC) (Vigneault and Émond, 1998). This handling process of horticultural produce is likely comparable to handling any other industrial products such as a TV set or a microwave oven. The main differences are in: the specificity of the environment due to the fact that the horticultural produce is still alive and requires special environmental conditions; the fragility of the produce to mechanical damage and its sensibility to the variations of the environmental conditions; and the length of the time the robot is used for. The particularities of environmental conditions in which the produce are handled and stored require the robot to be able to operate properly at ambient as well as in cold temperatures (close to $0°C$) and very high relative humidity (close to 100%), as well as in gas compositions that are hazardous to humans (close to 1% O_2 and up to 20% CO_2) in which the produce is sometimes stored (Leblanc and Vigneault, 2006); as well in a continuous alternating conditions from ambient to any other conditions. In terms of fragility of the produce to mechanical damages, it is easy to understand that horticultural produce are much more sensitive than any industrial products made from metal, plastic or wood. In terms of period of usage, the industry robots are generally used for 12 months/year period, which makes their uses generally economically interesting. However for the horticultural application, each robotic application is of any duration from a few days to the entire year depending on the type of work the robot accomplishes, the produce the robot processes and the produce production period which is produce and localization dependent, and the production system (greenhouse *vs* in-field production, fresh *vs* stored produce, cold climate *vs* subtropical, climacteric *vs* non-climacteric produce, etc). All these horticultural produce particularities must be considered during the design of any robot designated to horticultural applications.

2. Handling of fresh horticultural produce

Postharvest technologies include the objectives of maintaining the fresh quality of the produce in terms of appearance, texture, flavor, nutritive value, etc, protecting produce, maintaining food safety, and reducing the average losses between harvest and consumption (Saraswathy *et al.*, 2010). The time a produce is exposed to any adverse condition is generally directly proportional to the decrease of quality of any horticultural produce (Leblanc and Vigneault, 2008). Hence, the processing time is one of the most important and controllable factors affecting the quality of the produce. Any delay will generate soon or later important effects on the final quality of the produce (Vigneault *et al.*, 2004 & de Castro *et al.*, 2005) at the marketing step, which may result in a significant loss of marketable value (Leblanc and Vigneault, 2008).

Another controllable factor affecting the quality of the produce is the number of times a product is manipulated or in contact with any fixed or mobile object without being protected (Vigneault *et al.*, 2002 & Vigneault *et al.*, 2009). Packaging systems should be designed for rapid and efficient cooling (Vigneault *et al.*, 2002; Vigneault *et al.*, 2004b & de Castro *et al.*, 2005b) to prevent physical damage to the produce and to ease handling and storing processes (Vigneault *et al.*, 2009). Finally, storage conditions of refrigerated and controlled atmosphere should be chosen and maintained to retard the deterioration of the perishable produce due to: (1) aging, ripening, softening and color changes (Hoenht *et al.*, 2009); (2) undesirable metabolic changes and respiratory heat production(Leblanc and Vigneault, 2008) ; (3) moisture loss and the resulting wilting (Alvo *et al.*, 2004); (4) spoilage by invasion of bacteria, fungi or yeasts (Kader, 2002); and (5) undesirable physiological processes such as sprouting (Bachmann, 2010).

2.1 Automation and robots for handling of fresh horticultural produce

To maintain the quality of horticultural produce by applying prior knowledge, manufacturers supply a range of solutions based on control, automation and robotics. For example, to reduce processing time a produce is exposed to any adverse conditions, the handling industry proposes to reduce the time between harvest and cold storage by using high speed processing systems such as display container loaders for potato, onion, carrot and other pre-packs, which are capable of obtaining a processing rate up to 80 units of 2.5 kg per minute (Abar Automation, 2011). The automated loaders are generally available for loading 800 × 600 mm footprint containers with options available for the 600 × 400 mm standard RPCs (Abar Automation, 2011). In this process, as all the main movements including package rotation are already carried out in the other industries by a robot, very few conventional technology and mechanical components are required, resulting in a very reliable system. Crate loaders are also available for pre-pack picking and placing systems that load bags, clamshell and other small package units into the 600 × 400 mm North America and Europe standard crates (Vigneault *et al.*, 2009), 400 × 300 mm plastic crates or cases at a processing rate of more than one unit per second (Abar automation, 2009). The robot is housed with a compact space saving structure and space costs, and are fed with bulk or packed produce feeding conveyors. Outer types of packing units such as wooden crates, cardboard boxes, or RPCs are supplied to the system through conveyors either from remote or low level crate de-stackers. The remote de-stacker located on a mezzanine floor or adjacent to the pick-and-place robot separates the packing units from the top of the stack and delivers them over the packing lines. Robot de-stackers and palletizers provide a flexible and relatively low cost solution and high throughput with two or more action spots for single or multiple production lines of pallet loading.

During the picking and placing process in a packinghouse, the multitask operations allow bags, cases, boxes, RPCs and even layer sheets to be placed by a single robotic installation (Abar automation, 2009). Picking and placing is where a robot moves items from one place to another, for example from a conveyor in a production line to another in a packaging line.

The robot is often guided by a vision system (Piasek, 2010). Pick and place robot cells (Fig. 1) are available for pre-packed bags or trays containing fresh or processed vegetables, salads or mushrooms ranging in weights from a few grams to 50 kg, and are capable of loading plastic crates and roll containers (Abar automation, 2011 & Robots & Robot Controllers Portal, 2010).

Fig. 1. Pack rotation carry out by an ultra reliable (Courtesy: Abar automation, 2009).

The robotic work cells are becoming more complex. From de-moulding and assembling parts for the car industry to packaging complex objects, the robots are being asked to do faster and more complex moves. This ability to accomplish more sophisticated tasks lies in the capacity to develop more flexible end-of-arm end effecters. A customizable robot has the ability to increase functionality and repeatability as well as allow a higher percentage of automation function within the cell (KUKA Industrial Robots, 2010). For example, the crate sorting system at the Hofbräuhaus Brewery had to deal with different types of beer crates, meaning that the level of mixed packages in the system rose and manual handling became less cost-effective (Raghavan *et al.*, 2005). In addition, the brewery wanted to spare its employees from the heavy physical work of palletizing 2,500 to 3,000 crates per day. The new solution can also be integrated into the existing equipment. The brewery is looking for a customized automation concept. Finally, the system had to be installed in a small space, while at the same time being easily accessible (Raghavan *et al.*, 2005). This brewery example is quite similar in complexity to horticultural packinghouse where robots have to at the same time handle multiple types of fragile and light produce from conveyer to boxes, and heavy boxes to palette with care and precision, with a short time to maintain high production rate. Although the similarity in the complexity of the tasks to be accomplished is high which demonstrates the possibility of developing robots for horticulture application, the technologies are not directly applicable to horticultural produce with so many variations. Much modification to the industrial robots is required to adapt these robots for horticultural purposes. On April of this year was presented (Zheng, 2011) in China an intelligent robot (Fig. 2) ables to examine growing conditions such as temperature, humidity, light, detect disease of the vegetables and pick up the ripe ones through identifying the colour.

Fig. 2. A robot for vegetable planting (Courtesy: Xinhuanet, Zhu Zheng, 2011)

On the other hand, Dr. Hayashi Shigehiko in the the Bio-oriented Technology Research Advancement Institution (Japan) developed the harvesting robot of strawberry in oder to keep the high quality of strawberry freshness (Fig. 3).

Fig. 3. A harvesting robot of strawberry (Courtesy: BRAIN, Shigehiko Hayashi, 2011).

3. Storing of fresh horticultural produce

Various storage methods for fresh horticultural produce are used commercially (Liu, 1991 & Li, 2009). The most common and frequently used storage system is the refrigerated storage room (Raghavan *et al.*, 2005). This storage system consists of continuous and uniform control of temperature and humidity inside of well insulated rooms or buildings (Li, 2009). The ambient storage conditions for fresh horticultural produce vary in temperature from 0 to 15° C and in relative humidity from 65 to 98% [Kader, 2002; Raghavan *et al.*, 2005; Liu, 1991 & Li, 2009). These conditions are generally controlled by mechanical refrigeration [FAO, 2010 & Hollingum, 1999] and electric thermostat and humidistat, or using an electronic automated system based on adjusted enthalpy of the ambient air (Markarian *et al.*, 2006).

The controlled atmosphere (CA) storage frequently used in postharvest technology of fruits and vegetables consists of controls on the concentrations of O_2, CO_2 and N_2, within a perfectly sealed refrigerated storage room [38]. Modified atmosphere (MA) storage or packaging (MAP) is also used for horticultural produce. It consists generally of modifying gas composition of the internal atmosphere of a package unit that could be as small as 200 g (eg, for leafy vegetable) or a large transportable unit such as a pallet size of 250 to 1,000 kg (eg, for strawberry). The gas concentration encountered in CA and MAP are not different generally except the size of the units which is extremely different and use completely different atmosphere generation and controlling systems. The O_2 gas concentration used for CA and MAP is from 21% (ambient air) to 0%, while the CO_2

concentration increases from 0.003% to values sometimes reaching 15 to 20% de-pending on the produce stored. The N_2 is normally used as a filling gas for both MAP and CA technologies (Raghavan *et al.*, 2010). The ethylene (C_2H_4) is sometimes controlled by scrubbing such as potassium permanganate or O_3 (ozone), or by flushing for better quality control of produce responding to it. The control of temperature, humidity and atmospheric composition may be performed manually, but automated systems frequently measuring the gas compositions and performing the adequate adjustment of the gas processing system also exist to achieve more accurate and uniform conditions (Raghavan *et al.*, 2010).

3.1 Robots for storing of fresh horticultural produce

For storage of fresh produce, the robot handling system must be able to operate within the ambient conditions generally encountered in the storage rooms. At the same time, the robots have to be able to work within the conditions encountered outside of the storage room since they are continuously required to handle the produce to and from the storage room. They must then be resistant to heavy ambient humidity condensation due to this continuously changing environment. Besides the humidity condition, the basic principles used for developing industrial storage robots are basically the same for horticultural purposes. The robot must be able to circulate in a specific environment recognising fixed and moving objects consisting of rows and columns of produce, walls, doors, humans and other robots working individually or in collaboration with human and other robots.

There are two great advantages of robots used for horticultural storage purposes. The first is the robot's capacity for managing extremely large quantities of information associated with any specific produce. These pieces of information could be related to the type of produce, the origin, the harvesting date, the processing date, the initial quality, and its eventual destination, etc (Leblanc and Vigneault ,2006). The information could be easily transferred from and to the robots for managing purposes and improvement of efficiency. For example, maintaining the link between the information and the produce is essential for choosing and executing any managing types such as "the first in first out", "consumer preference", "quality based" or "shelf live" objectives. The second advantage is that in a great facility, a robot can be considered as an economical aspect in the handling the produce such as optimisation of space occupancy, organisation of traffic circulation and managing picking order (Piasek, 2010) to maximise handling efficiency and minimise produce waiting time on docks or truck loading process duration.

Another advantage of a robot, which has not been explored yet in horticulture, is the robot's capacity to work in hazardous environment such as CA room or high O_3 concentration atmosphere for C_2H_4 control. In the case of C_2H_4 control, one has to understand that the optimal O_3 level for the best control of C_2H_4 is generally too high for worker safety, which considerably limits the use of this interesting technique. For a CA room, the main problem is the fact that the quality of the produce is highly related to the time delay between harvest and storage and between storage and commercialization. It is frequent to encounter produce that has been harvested at the optimum maturity level but has to wait for days and even weeks before being under CA due to the time required to fill the CA room. The same way, it is very common to encounter CA rooms that were degassed for marketing a first load of produce and had to wait weeks and months before the last load quits the room. An airlock system has been developed for CA apple storage which can maintain the gas composition inside of the room continuously (Vigneault, 2009), reduce the waiting time before and after storage to a few hours only. However, workers have to use autonomous breading systems to enter in the airlock and transfer the produce from and to the CA room, which has to be performed with a very high precaution level. Human life can be in danger if any mistake happens (Vigneault, 2009).

4. Transportation of fresh horticultural produce

Ground transportation of fresh horticultural produce is usually by trucks and occasionally by railway (Leblanc and Hui, 2005). Overseas transportation is by ship (Tanner and Smale,2005) or plane (Thompson et al., 2004). A limited amount of high-valued produce is sometimes transported overland by air (Vigneault et al., 2009). The basic requirements for conditions during transportation are similar to those needed for storage (Vigneault, 2005), including proper control of temperature and humidity, and adequate ventilation (Hui et al., 2006 & Hui et al., 2008a). In addition to ambient condition require-ment, the produce must be immobilised by proper packaging, stacking, and bracing methods to avoid excessive movement or vibrations (Vigneault et al., 2009) without being detrimental to air circulation to maintain a uniform temperature distribution within a load (Hui et al., 2008b).

4.1 Robots for transportation of fresh horticultural produce

Loading trucks with produce can be done robots with a vision machine and integrated sensors able to detect objects or people obstructing their way and find and follow a specific path (Belforte et al., 2006). For the purpose of this handling associated with transportation, the robots should be compact enough to perform this task within a restricted work area (Garcia Ruiz et al., 2007), robust but able to perform delicate and very precise operations since the load placement within the truck greatly affects the air circulation (Vigneault , 2005; Hui et al., 2006 & Hui et al., 2008a). To perform these tasks, transportation robots coupled with forks and elevator can carry packed industrial products as well as horticultural produce (Logistic Systems Design, 2010).

5. Conclusion

Packaging and refrigeration prevent physical damage and decrease the physiological and pathological damages to the produce. Automation and robots are capable of speeding up the postharvest processes, decrease the period during which the perishable produce is exposed to undesirable conditions, and enhancing the facility and the safety level of human tasks for handling fresh horticultural produce. Robot palletisers could provide more than one course at the time and work within multi line production. Picking and placing robot cells adapted from industrial uses for pre-packed bags, trays or boxes containing fresh fruits or vegetables are becoming more available and able to accomplish complex tasks, but they have to be redesigned to meet with the specific require-ments and environmental conditions of horticulture.

6. Acknowledgment

The authors would like to thank to the Mexican, Canadian and Japanese Ministries of Education for the financial support. Moreover, we thank the robot companies and institutions for the photographs.

7. References

Kitamura S.; Oka K and Takeda F. (2004). Development of Picking Robot in Greenhouse Horticulture. *SICE Annual Conference*, pp. 3176–3179, ISBN 0780320247 Okayama, Japan

Wang M. What will modern agriculture bring to people? Sciences by Experts, Series 3. (2009) *China Publishing Group*, ISBN/ISSN 9787500084785 pp. 145– 160, Beijing, China

Sarig Y, Thompson JF and Brown GK. (2010). Alternatives to Immigrant Labor? The Status of Fruit and Vegetable Harvest Mechanization in the United States. 2000: 29.07.2010. Available from http://www.cis.org/articles/2000/back1200.html

ElMasry G, Nassar A, Wang N and Vigneault C. (2008) Spectral methods for measuring quality changes of fresh fruits and vegetables. *Stewart Postharvest Review*, Vol.4 , No. 4, pp. 1–3.13, ISSN 1745-9656

Li Z, Wang N and Vigneault C. (2006) Electronic nose and electronic tongue in food production and processing. *Stewart Postharvest Review*, Vol.2, No.4, pp. 7.1–7.6., ISSN 1745-9656

Lino ACL, Sanchez J and Dalfabro I. M. (2006). Image processing techniques for lemons and tomatoes classification. *Bragantia*, Vol.3, pp. 85–789, ISSN 0006-8705

Chen M. (2008). Non-destructive Measurement of Tomato Quality using Visible and Near-infrared Reflectance Spectroscopy. Master degree thesis. G.S.V. Raghavan and C. Vigneault (Supervisors). *Department of Bioresource Engineering. Macdonald Campus, McGill, University.* Sainte-Anne de Bellevue, Québec, Canada

Li Z, Wang N, Vigneault C and Raghavan GSV. (2009). Ripeness and rot evaluation of 'Tommy Atkins' mango fruit through volatiles detection. *Journal of Food Engineering,* Vol. 91, pp. 319–324. ISSN 0260-8774

ElMasry G, Wang N, Qiao J, Vigneault C and ElSayed A. (2008). Early detection of apple bruises on different background colors using hyperspectral imaging. LWT – *Food science and technology*, Vol. 41, No.2, pp, 337–345, ISSN 0023-6438

ElMasry G, Wang N and Vigneault C. (2009). Detecting Chilling Injury in "Red Delicious" Apple Using Hyperspectral Imaging and Neural Network. *Postharvest Biology and Technology,* Vol. 52, No.1, pp. 1–8, ISSN 0925-5214

Key Technology. Lettuce Core Removal System. (2010). 25.07.2010. Available from http://www.key.net/ products/oncore-lettuce-core-removal-system/default.html

Applied Sorting Technologies. (2010). 24.07.2010. Available from http://www.appliedsorting.com.au/ index.php?~iframe1

Vigneault C and Émond JP. (1998). Reusable container for the preservation of fresh fruits and vegetables. *Agriculture and Agro-Food Canada and Laval University.* United States Patent Office, Washington, Washington DC, USA

Leblanc D.I. and Vigneault C. (2006). Traceability of environmental conditions for maintaining horticultural produce quality. *Stewart Postharvest Review,*Vol.2 , No. 2, pp. 4.1–4.10, ISSN 1745-9656

Saraswathy S, Preethi TL, Balasubramanyan S, Suresh J, Revathy N and Natarajan S. (2010). Postharvest Management of Horticultural Crops. *Agrobios*, India

Leblanc DI and Vigneault C. (2008). Predicting quality changes of fresh fruits and vegetables during postharvest handling and distribution. *Stewart Postharvest Review*,Vol.4 , No. 4, pp. 1.1–1.3., ISSN 1745-9656

Vigneault C, Gariépy Y, Roussel D and Goyette B. (2004a). The effect of precooling delay on the quality of stored sweet corn. *International Journal of Food, Agriculture and Environment*,Vol.2, No. 2, pp. 71–73. ISSN (printed) 1459-0255. ISSN (electronic) 1459-0263

de Castro LR, Vigneault C, Charles MT and Cortez LAB. (2005a). Effect of cool-ing delay and cold-chain breakage on Santa Clara tomato. *International Journal of Food, Agriculture and Environment* Vol.3 , No. 1, pp. 49–54. ISSN (printed) 1459-0255. ISSN (electronic) 1459-0263

Vigneault C, Bordint MR and Abrahão RF. (2002). Embalagem para frutas e hortaliças. In : Cortez LAB, Honório SL and Moretti CL (Eds). Resfriamento de frutas e hortaliças. *Embrapa Informaçã Tecnológica*, Brasília, DF, Brazil, ISSN 1678-3921

Vigneault C, Thompson J and Wu S. (2009). Designing container for handling fresh horticultural produce. In Benkeblia N (ed). *Postharvest Technologies for Horticultural Crops*. Research Signpost, India, Vol. 2, pp. 25–47. ISBN 978-81-308-0356-2

Vigneault C, Markarian NR, da Silva A and Goyette B. (2004b).Pressure drop during forced-air ventilation of various horticultural produce in containers with different opening configuration. *Transaction of the ASAE*, Vol.47, No.3, pp. 807–814, ISSN 0001-2351

de Castro LR, Vigneault C and Cortez LAB. (2005b). Effect of container openings and airflow on energy required for forced air cooling of horticultural produce. *Canadian Biosystems Engineering*, Vol.47, No., pp. 3.1–3.9., ISSN 1492-9058

Hoehn E, Prange R and Vigneault C. (2009). Storage technology and applications. In: Yahia EM (ed). Modified and controlled atmospheres for the storage, transportation, and packaging of horticultural commodities. *CRC Press (Taylor & Francis Group)*, pp. 17–50, ISBN 9781420069570.ISBN10 1420069578

Alvo P, Vigneault C, DeEll JR and Gariépy Y. (2004). Texture characteristics of carrots as affected by storage temperature and duration. *International Journal of Food, Agriculture and Environment* 2004: 2(2): 33–37, ISBN 9781420069570

Kader A (ed). (2002). Postharvest technology of horticultural crops. *University of California*, 2002: Publication 3311. Davis, CA, USA

Bachmann J and Earles R. (2010). Postharvest Handling of Fruits and Vegetables. 27.07.2010. Available from http://attra.ncat.org/attra-pub/postharvest.html.

Abar automation. (2011). Pick and Place Robots for Fresh Produce, Netherlands. 18.10.2011. Available from http://www.abar.nl/pick_and_place_fresh_produce.htm

Piasek D. (2010). Order Picking: Methods and Equipment for Piece Pick, Case Pick, and Pallet Pick Operations. *Logistics Handling, Dematic*, Le Dôme, Luxembourg. 28.07.2010.Available from
http://www.logisticshandling.com/absolutenm/templates/article-conveyors_sortation.aspx?articleid=36&zoneid=29

Robots & Robot Controllers Portal. (2010). Specify Multi-functioning End-of-Arm Tooling for Robotic Workcells. 26.07.2010. Available from http://www.automation.com/robots-robot-controllers/preview-articles/more

KUKA Industrial Robots. (2010). Handling of beverage crates with 180 PA robot. 25.07.2010. Available from http://www.kuka-robotics.com/en/solutions/solutions_search/L_R171_Handling_of_beverage_crates_with_180_PA_robot.htm.

Raghavan GSV, Vigneault C, Gariépy Y, Markarian NR and Alvo P. (2005). Refrigerated and controlled/modified atmosphere storage. In Barrett D, Somogyi L and Ramaswamy H (eds). *Processing Fruits, Science and technology. 2nd edition. CRC Press.* ISBN 9781420069570 ISBN10-1420069578

Zheng Zhu (2011). Robots for vegetable planting works at 12th China Int'l Vegetable Sci-tech Fair in Jinan, China's Shandong. 20.04.2011. Available from http://news.xinhuanet.com/english2010/photo/2011-04/20/c_13838241_2.htm

Shigehiko H. (2011). Specific Project Research Team (Robot) Bio-oriented Technology Research Advancement Institution. Japan. 30.11.2011 Available from http://brain.naro.affrc.go.jp/iam/Team/iam_team_robo00.html

Liu FW. (1991). Precooling of horticultural crops. In: Yahia EM and Higuera IHC. (eds.) Memorias Simposio Nacional Fisiologia y Tecnologia Post-cosecha de Productos Horticolas en Mexico. *Noriega Editores*, Mexico, pp.235–240.

Li FW. (2009). Postharvest Handling in Asia 2. Horticultural Crops Food and Fertilizer. 21.12.2009. Available from http://www.agnet.org/library/eb/465b/

FAO. (2010). Small-scale post-harvest handling practices. Section 6: Temperature and relative humidity control. 27.07.2010. Available from http://www.fao.org/WAIRdocs/x5403e/x5403e08.htm.

Hollingum J. (1999). Robots in agriculture. *Industrial Robot: An International Journal*, Vol. 26, No. 6, pp. 438–446, ISSN 0143-991X

Markarian NR, Landry JA and Vigneault C. (2006). Development of a model for simulating ambient conditions in fresh fruit and vegetable storage facility. *International Journal of Food, Agriculture and Environment*, Vol.4, No.1, pp. 34–40, ISSN (printed) 1459-0255. ISSN (electronic) 1459-0263

Raghavan GSV, Gariépy Y and Vigneault C. (2010). Controlled Atmosphere Storage. *Encyclopedia of Agriculture, Food, and Biological Engineering, 2nd Edition. Taylor & Francis* 2010: (In press).

Vigneault C. (2009). La combinaison d'un sas AC et un système de refroidisse-ment à l'eau pour une meilleure gestion de la qualité des pommes. *Agriculture and Agri-Food Canada*, Saint-Jean-sur-Richelieu, Québec, Canada, 3pp

LeBlanc DI and Hui KPC. (2005). Land transportation of fresh fruits and vegetables: An update. *Stewart Postharvest Review*, pp. 1.4.1–1.4.13. ISSN 1745-9656

Tanner D and Smale N. (2005). Sea transportation of fruits and vegetables: An update. *Stewart Postharvest Review*,pp. 1.1.1–1.1.9, ISSN 1745-9656

Thompson JF, Bishop CFH and Brecht PE. (2004). Air transport of perishable products. *Agriculture and Natural Resources*, University of California, 2004: Publication 21618. 22 pp. Davis, CA, USA

Vigneault C, Thompson J, Wu S, Hui KPC and LeBlanc DL. (2009).Transportation of fresh horticultural produce. In Benkeblia N (ed). *Postharvest Technologies for Horticultural Crops. Research Signpost*, Kerale, India. Vol. 2, pp. 1–24, ISBN 978-81-308-0356-2

Vigneault C. (2005). Transport of fruits and vegetables. *Stewart Postharvest Review*,Vol.1, No.1, pp. 6.1–6.4, ISSN 1745-9656

Hui KPC, Raghavan GSV, Vigneault C and de Castro LR. (2006) Evaluation of the air circulation uniformity inside refrigerated semi-trailer transporting fresh horticultural produce. *Journal of Food, Agriculture and Environment* ,Vol.4, No. 1, pp. 109–114, ISSN 1459-0255

Hui KPC, Vigneault C, de Castro LR and Raghavan GSV. (2008a). Effect of different accessories on airflow pattern inside refrigerated semi-trailers transporting fresh produce. *Applied Engineering in Agriculture*, Vol. 24, No. 3, pp. 337–343, ISSN 0883-8542

Hui KPC, Vigneault C, Sotocinal SA, de Castro LR and Raghavan GSV. (2008b). Effects of loading and air bag bracing patterns on correlated relative air distribution inside refrigerated semi-trailers transporting fresh horticultural produce. *Canadian Biosystems Engineering*, pp. 3.27–3.35, ISSN 1492-9058

Belforte G, Deboli R, Gay P, Piccarolo P and Ricauda Aimonino D. (2006). Robot Design and Testing for Greenhouse Applications. *Biosystems Engineering*,Vol. 95, No. 3, pp. 309–321,ISSN 1537-5110

García Ruiz MA, Gutiérrez Díaz S, López Hernández HC, Rivera Ibáñez S and Ruiz Tadeo AC. (2007). Estado del arte de la tecnología de robots aplicada a invernaderos state of the art of robot technology applied to greenhouses. *Avances de Investigación Agropecuaria*, Universidad de Colima, México, Vol. 11, pp. 53–61.

Logistic Systems Design. (2010). Handling Systems. 21.07.2010. Available from http:/www2.isye.gatech.edu/ ~mgoetsch/cali/logistics_systems_design/material_handling_systems/ material_handling_systems.pdf.

9

Sustentable Use of the Wetting Agents in Protected Horticulture

Carlos Guillén and Miguel Urrestarazu
Universidad of Almería,
Spain

1. Introduction

The wetting agents in agriculture are materials that are classified within a larger group identified as adjunvants (Cahn and Lynn, 2000; Hazen, 2000; Kosswig, 2000; Thacker, 2003; Young, 2003; Lynn y Bory, 2005). According to their chemical nature, adjuvants are grouped into four families: surfactants, oils, inorganic salts and non-traditional adjuvants (figure 1). Nowdays some of the wetting agentes can be considered inside of called Green Chemistry (Carrasco and Urrestarazu, 2010).

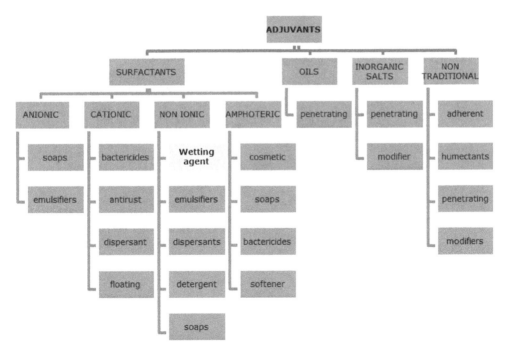

Fig. 1. Classification of adjunvants according to their chemical nature and function

1.1 Surfactant

A common mistake is the use of adjuvantand surfactant terms inter changeably, which arises from the same discrepancy between producers(Young, 2003). The tru this that surfactantisa chemical family of adjuvants that improve emulsion, dispersion, foam, wet or other properties of a fluid to alter the characteristics of surface or interface and surface tension (Hazen, 2000, Thacker, 2003; Young, 2003). The termsurfactantis derived from the contraction of three words Surface Active Agents (Thacker, 2003; Rosen, 2004, RAE, 2001).

Global consumption of surfactants in agriculture is 250 thousand tons per year, 180 thousand of which are incorporated in the formulation of fitosanitariosy 70 000 tonnes are used as tank mix adjuvants (Thacker, 2003). Besides agriculture surfactants are used in various products such as motor oils for cars, pharmaceuticals, in detergents, in materials used in oil exploration and floating agents in mining, in recent decades has extended its application to electronic printing, magnetic recording, biotechnology and viral research (Rosen, 2004).

Surfactant are molecules of low to middle rate molecular weight, it is amphipathic nature, it is containing a hydrophobic or lipophilic part (carbon chain) and a hydrophilic or lipophobic (Malmnsten, 2002, Rosen, 2004; Tadros, 2005).

The hydrophobic groups are alkyl and alkylaryl groups mostly hydrocarbon, hydrophilic groups can be ionic or nonionic (Kosswig, 2000). According to this type ionic surfactants may be anionic, cationic, amphoteric and nonionic surfactants (Figure 2). In an anionic surfactant hydrophilic segment of the molecule forms an anion when dissolved in water. The opposite occurs in the cationic surfactant in which the active portion of the molecule in the hydrophilic segment is only a cation when dissolved in water. An amphoteric surfactant is capable of forming anions in aqueous solution depending on pH or cations. The nonionic surfactant is a surface active agent without ionic polar group is not ionized in aqueous solution (American Society for Testing and Materials E 1519, 1999, Cahn and Lynn, 2000; Kosswig, 2000; Malmnsten, 2002; Rosen, 2004, Young, 2003, Lynn and Bory, 2005).

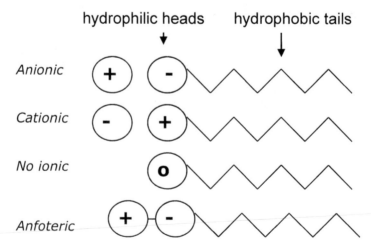

Fig. 2. Diagram of a humectant molecules

Reference	Physical properties of substrate				Growth and yield	Uptake of nutrients	Phytotoxicity
	Wettability	AV	WHC	EAW			
Sheldrake and Matkin (1971)							according surf.
Airhart et al. (1980)			increase				according dose
Wilson (1985)		increase					
Powell (1986)	increase	increase					according dose
Milks et al. (1989)		increase	decrease				
Bhat et al. (1990)					better post-harvest	increase Ca^{2+}	
Bhat et al. (1992)			increase				according plant
Handreck (1992)			increase				according dose
Handreck and Black (1994)			increase				
Elliot (1992)	increase		non effect				
Blodgett et al. (1993)	increase		non effect	increase			
Cid et al. (1993)					increase		
Blodgett et al. (1995)							accordingplant
Riviere et al. (1996)	increase		increase	increase			
Michel et al. (1997)	increase						
Reinikainen and Herranen (1997)	increase	increase					
Bilderback and Lorscheider (1997)		decrease	increase		increase		
Cid et al. (1998)			increase		increase	increase Ca^{2+}	
Guillen and Urrestarazu (2006)					increase	increase	
Urrestarazu et al. (2008)	increase		increase	increase	increase	increase	according dose

AV: air volume, WHC: water holding capacity, EAW: easily available water

Table 1. Summary of research on the effect of application of wetting agent in ornamental and horticultural cultures on substrate

The nonionic surfactants are chemically less active, but also less phytotoxic (Powell, 1986; Bures, 1997; Reinikainen and Herranen, 1997;) and less irritating than anionic and cationic (Malmnsten, 2002). They can with stand hard water and soluble in water and organic solvents, but can be sensitive to high temperatures (Rosen, 2004). Besides its critical concentration of micelle formation is twice lower than the anionic surfactants (Tadros, 2005). These nonionic surfactants are excellent wetting at very low concentrations (Thacker, 2003). They are the most dominant adjuvants of surfactants commercially available for the application of herbicides, they are less toxic to mammals (Young, 2003).

1.2 Wetting agent

Wetting agent is a substance used to reduce surface tension (Figure 3) and lead to better contact of a solution or suspension with a surface (WSSA Herbicide Handbook, 1994). Generally wetting agentis a surfactant, whose effectiveness is measured by the increase of spread of a liquid on a surface or the contact angle of liquid with the surface (Shurtleff and Averre, 1998).

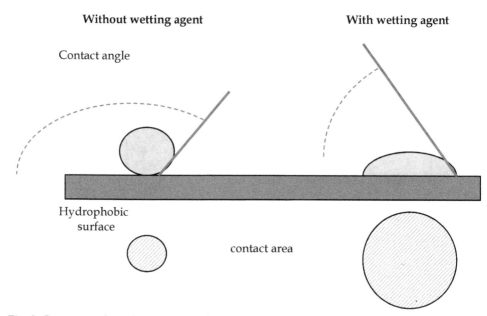

Fig. 3. Contact angle and contact area for a drop water with and without a wetting agent on a hydrophobic surface

The wetting agents have been successfully used to improve the uniformity of wetting, improve relationships air: water of organic substrates and to improve nutrient availability to plants, without altering the physical characteristics of the substrate (Powell, 1986; Baht et al., 1992, Cid et al., 1992, Blodgett et al., 1993).

Crabtree and Gilkes (1999) indicate that the wetting agent also apply water repellent soils and have several benefits for the soil-plant system because it improves the soil wet releases the fertilizer applied in the soil solution increases the mineralization of organic matter, wet consistency, improves water infiltration, reduces the flooded surface, reduces evaporation and increases the efficiency of water use in farming.

In trade, de Linan (2006), we found in the group of adjuvants to 30 products, among which are 17 active substances to wetting (Table 2).

	Material	Coadjuvant kind
1	Alkyl ethoxylated fatty amines 73.5% w/v. SL	Surfactant nonionic
2	Alkyl ethoxylated fatty amine / propoxylated 48.7% w/v. SL	Surfactant nonionic
3	Alquil aminas grasas etoxiladas/propoxiladas 92% p/v. EC	Surfactant nonionic
4	Alquilpoliglicol 20% w/v. EC	Nonionic adjuvant
5	Alquilpoliglicol 44% + 1.5% sodium dioctyl w/v. SL	Surfactant nonionic
6	Sodium Alquiletersulfato 29% w/v. SL	Surfactant nonionic
7	Ethylene Alquilfenilhidroxipolioxi + 10.1% synthetic latex 45.45% w/v. EC	Surfactant nonionic
8	Montan wax 20% w/v. EW	Adherent-adjuvant and non-ionic wetting
9	ammonium dodecylbenzenesulfonate 20% w/v. SL	Adjuvant anionic wetting power
10	Ammonium dodecylbenzenesulfonate 20% w/v. EC	Adjuvant anionic wetting power
11	Sodium dodecylbenzenesulfonate 5% + nonilfeniloxietilenatosulfonado 2% w/v. SL	With wetting power
12	Methyl oleate / methyl palmitate 34.5% w/v. EC	It reduces the surface tension
13	Surfactant nonionic 20% p/v. SL	Surfactant nonionic
14	EterNonilfenolpolietilenglicolr 20% p/v. SL	Surfactant nonionic
15	Eter Nonilfenolpolietilenglicol 30% + inorganic acid 30% w/v. SL	With wetting power
16	Copper phthalate, are grouped a series of copper salts of fatty acids a dozen	With wetting power
17	Copper phthalate 66.5% w/v. EC	Adjuvant anionic wetting power

Source: De Liñán (2006)
EC: Emulsifiable concentrate, EW: Emulsion oil-water; SL: soluble concentrate, p/v: weight/volume

Table 2. Wetting adjuvant for agricultural use in Spain

2. Application of wetting agent in crop production

2.1 Assessment of toxicity

It is important before using a new wetting agent, evaluate their toxicity to plants (Handreck and Black, 1994). The bioassays used to assess the phytotoxicity of substrates (Zucconi et al., 1981, Ortega et al., 1996) were used to evaluate the toxicity of some products used in soilless culture (Urrestarazu and Mazuela, 2005). Urrestarazu et al. (2006) reported an adequate concentration of a surfactant with bioassays of tomato and cress.

2.2 Improving substrates

2.2.1 Wettability and water holding of growing media

The wettability of growing media (figure 4) is an important characteristic that could be limiting due to alternations of humectation and desiccation or drying accidental events that could modify considerably and reversibly or not the properties of the substrate during growing crops (Da Silva et al., 1993; Otten et al., 1999, Chambers and Urrestarazu, 2004, Lemaire et al., 2005).

Reduction of wettability contributes to vertical flow and goes against horizontal flow and water retention in growing media (Handreck and Black, 1994; Beeson and Haydu, 1995; Dekker and Ritsema, 1994, 1996 and 2000; Salas and Urrestarazu, 2004; Dekker et al., 2005).

Greater water retention capacity at low matrix potential is very important for optimal plant growth (Plaut et al., 1973, Plaut and Zieslin, 1974, Feigin et al., 1988; Raviv et al., 2002, Sahin et al., 2002, Chambers and Urrea, 2004). There have been several experiments to correlate the growth of plants with water retention and air capacity of the substrates and these has been found that the plant qrowth is highly correlated with water retention and low air capacity is associated with a low crop growth (Allaire et al., 1996). These problems of hydrophobia and low holding water in substrates can be overcome by application of wetting agent.

new peat   new coir waste   used coir waste

Fig. 4. Water repellency of dry organic substates

It has been shown that surfactants facilitate the movement of water into and through the substrate, control the infiltration of water distribution and drainage, because it affects reserves of moisture, nutrient availability and aeration, with an optimal concentration of these surfactants improve rewetting potential of substrates and reduce root stress related problems, allowing greater control of plant growth (Powell, 1986, Bhat et al., 1990).

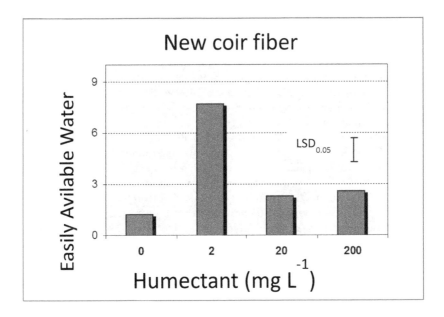

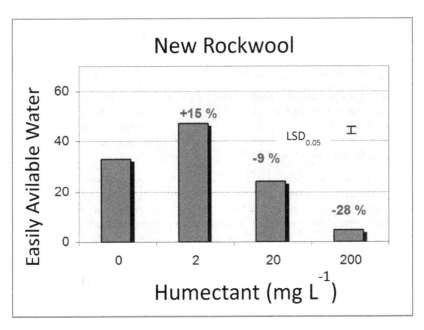

Fig. 5. The square of Humectant delete de red line behind, this must be like de red line as below

2.2.2 Improving physical properties of substrates

There is a considerable amount of articles about application of surfactants in ornamental and horticultural crops on different substrates. Table1 shows some research on application of wetting agent on crop son substrate. Surfactants applied to organic growing media such as peat and pine fiber can improve absorption of water (wettability) of substrate (Handreck, 1992, Elliot, 1992; Blodgett et al., 1993, Riviere et al., 1996; Michelet al., 1997; Reinikainen and Herranen, 1997), also can increase air capacity, raising the drain (Wilson, 1985; Powell, 1986; Milksetal, 1989; Reinikainen and Herranen, 1997) or can improve the water holding capacity (Airhart et al., 1980; Handreck, 1992; Handreck and Black,1994, Elliot, 1992; Bilderback and Lorscheider, 1997, Cid et al., 1998), increasing the water held inmicropores. May also increase the water content readily available, without increasing the capacity of container (Blodgett et al., 1993) or increase the total available water content with increased water holding capacity (Riviere et al., 1996.). Urrestarazu et al (2008) reported that the effect of wetting agent added through fertigation is directly dependant on the substrate type evaluated and it can reduce the available water and increase the easily avilable water and total water holding capacity.

2.3 Increasing efficiency in ornamentals

Also can improve growth and production of ornamental species if applied at rates not phytotoxic (Bhat et al., 1990, Cid et al., 1993; Bilderback and Lorscheider, 1997, Cid et al., 1998), even allow to increase the availability and absorption of nutrients, especially calcium absorption (Bath et al., 1990, Cid et al., 1998). And apparently the application in fertigation could be the best method of application, less phytotoxic than when initially mixed with the substrate (Bhat et al., 1990).

2.4 Increasing efficiency in vegetables

In application by fertigation, the surfactant increased nutrients uptake, especially potassium ($P = 0.08$), nitrate ($P = 0.15$) and phosphate ($P = 0.25$), in melon crop on coir waste. Also it was observed a decrease in the percentage of drainage and reducing the emission of nitrates, phosphates, potassium and calcium ($P \leq 0.2$) in melon crop, showing its usefulness in lower environmental pollution (Guillen and Urrestarazu, 2006).

This observed in coir mojantesmejoran shows that the properties of the substrate water quality by reducing the hydrophobic effect acquired especially for use. But we have discussed in the previous section that this property is related to the concentration of mojanteque used, hence the application by fertigation should be adjusted to find the desired effect. In short, the goal is to improve efficiency, since the increased incorporation counterpart brings in reducing polluting emissions.

In tomato trial, the first year was observed higher yield in crop on reused coir waste and greater efficiency in water use ($P < 0.10$), proving useful for improving the use of water when a wetting agent is applied by fertigation. Also was observed a greater efficiency in the second year in reused coir waste ($P = 0.07$) and reused rockwool ($P = 0.03$) (Guillen and Urrestarazu, 2006).

It is inferred that the wetting by fertigation is a useful tool to adjust the water relations of sutratos used, which is reflected in higher production volume of water consumed.

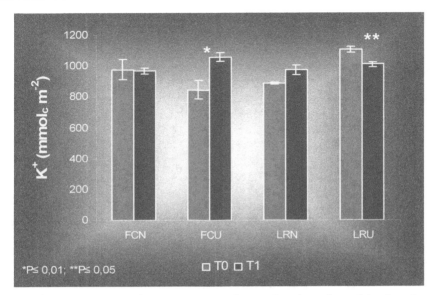

Fig. 6. Significant increases in the incorporation of potassium in melon cultivation of coconut fiber used. T0 = Control, T1 with humectant. FCN: New coir fiber, FCU: Used fiber coir. LRN: New rockwool, FRU: Used rockwool.

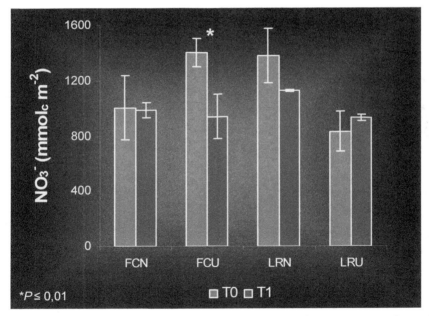

Fig. 7. Significant reduction in the emission of nutrients in a melon crop in reused coco fiber as a result of a wetting agent applied in fertigation. T0 = Control, T1 with humectant. FCN: New coir fiber, FCU: Used fiber coir. LRN: New rockwool, FRU: Used rockwool.

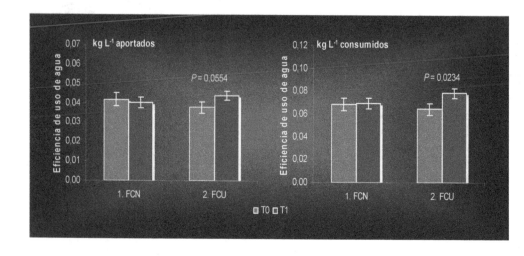

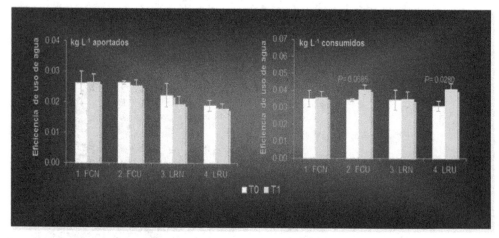

Fig. 8. Efficient water use in different substrates with and without wetting. T0= Testigo, T1 with humectant. FCN: New coir fiber, FCU: Used fiber coir. LRN: New rockwool, FRU: Used rockwool.

3. Acknowledgements

The authors acknowledge the collaboration of Ministry of Science and Innovation Spanish through the project FEDER AGL2010-18391.

4. References

Airhart, D.L.; Natarella, N.J. & Pokorny, F.A., (1980). Wetting a milled pine bark potting medium with surfactants.Forest Products Journal 30 (12), 30-33.

Allaire, S.; Caron, J.; Dúchense, I., Parent, L. & Rioux, J.A., (1996). Air filled porosity, gas relative diffusivity, and tortuosity, indices of prunasxcistena sp. Growth in peat substrates. Journal of the American Society for Horticultural Science 121(2), 236-242.

American Society for Testing and Materials E 1519, (1999). Annual Book of ASTM Standards, Vol. 11.05. Designation E 1519-95, Standard Terminology Relating to Agricultural Tank Mix Adjuvants. pp. 905-906.

Beeson, R.C. & Haydu, J. (1995). Cyclic microirrigation in container-grown land-scape plants improves plant growth and water conservation. Journal of Environmental Horticultura 13, 6-11.

Bhat, N.R.; Tayama, H.K.; Prince, T.L., Prince, T.A. & Carver, S.A., (1990). Effect of Aqua-Gro wetting agent on the growth, flowering, and postproduction quality of potted chrysanthemum, Ohio Florists' Association Bulletin 723, 3-6.

Bhat, N.R.; Prince, T.L.; Tayama, H.K. & Carver, S.A. (1992). Rooted cutting establishment in media containing wetting agent.HotScience 27(1), 78.

Bilderback, T.E. & Lorscheider, M.R. (1997). Wetting agents used in container substrates are they BMP'S?. ActaHorticulturae 450, 313-319.

Blodgett, A.M.; Beattie, D.J.; White, J.W. & Elliot, G.C. (1993). Hydrophilic polymers and wetting agents affect absortion and evaporative water loss. HortScience 28(6), 633-635.

Burés, S. (1997). Substratos. Ediciones Agrotécnicas, Madrid, pp. 342. ISBN Madrid, Spain. 342 pg. ISBN 84-87480-75-6

Cahn, A. & Lynn, J.L. (2000). Surfactants and detersive systems. Kirk-Othmer Encyclopedia of Chemical Technology, pp. 1142-1146.

Carrasco G. & Urrestarazu M. (2010). Green Chemistry in Protected Horticulture: The Use of Peroxyacetic Acid as a Sustainable Strategy. International Journal of Molecular Sciences, 11(5), 1999-2009.

Cid, M.C.; Socorro, A.R. & Pérez, R.L. (1993). Root growth and quality rating of Schefflera "Golden Capella" and Ficus "Starlight" on several peat-based substrates. Acta Horticulturae 342, 307-311.

Cid, M.C.; Muñoz, C.R.; Socorro, A.R. and Gonzáles, T.G. (1998). Wetting agent effects on peat properties related to nutrient solution losses and plant growth. Acta Horticulturae 458, 161-165.

Crabtree, W. & Gilkes, R. (1999). Banded wetting agent and compaction improve barley production on a water-repellent sand. Agronomy Journal 91, 463-467.

Da Silva, F.F.; Wallach, R. & Chen, Y. (1993). Hydraulic properties of sphagnum peat moss and tuff (scoria) and their potential effects on water availability. Plant Soil 154, 119-126.

Dekker, L.W. & Ritsema C.J. (1994). How water moves in a water repellent sandy soil 1, potencial and actual water repellency. Water Resources Research 30, 2507-2517.

Dekker, L.W. & Ritsema, C.J. (1996). Preferential flow paths in a water repellent clay soil with grass cover. Water Resources Research 32(5), 1239-1249.

Dekker L.W. & Ritsema, C.J. (2000). Wetting patterns and moisture variability in water repellent Dutch soils. Journal of Hydrology 231-232, 148-164.

Dekker, L.; Oostinde, K. & Ritsema, C. (2005). Exponential increase of publications related to oil water repellency. AustralianJournal of SoilResearch 43, 403-441.

De Liñán, C. (2005), Vademécum de productos fitosanitarios y nutricionales, 21ª ed. Ediciones Agrotécnicas S.L., pp 286-288. ISBN 84-87480-37-3, Madrid, Spain.

Elliot, G.C. (1992). Imbibition of water by rookwool-peat container media amended with hydrophilic gel or wetting agent. Journal of the American Society for Horticultural Science 117(5), 757-761.

Elliot, G.C. (1992). Imbibition of water by rookwool-peat container media amended with hydrophilic gel or wetting agent. Journal of the American Society for Horticultural Science 117(5), 757-761.

Feigin, A.; Gammore, R. & Gilead, S. (1988). Response of rose plants to NO_3 or Cl salinity under different CO_2 atmosphere.Soilless Culture IV, 66.

Handreck, K.A. & Black, N.D. (1994). Growing Media for Ornamental Plants and Turf.New South WalesUniversity Press, pp. 401. ISBN 0-86840-333-4, Kensington, Australia

Handreck, K. (1992). Wetting agents, how to achieve the best results. Australian Horticulture, August.

Hazen, J.L. (2000). Adjuvants-Terminology, Classification, and Chemistry. Weed Technology 14, 773-784.

Kosswig, K. (2000). Surfactants. Ullmann`s Encyclopedi of Industrial Chemistry, pp. 293-368.

Lemaire, F.; Dartigues, A.; Riviere, L.M.; Charpentier, S. & Morel, P. (2005). Cultivos en macetas y contenedores principios agronómicos y aplicación. INRA – MundiPrensa, pp. 210. ISBN: 84-8476-201-7, Madrid, Spain.

Lynn, J.L. & Bory, B.H. (2005). Surfactants. Van NostrandsEnciclopedia of Chemistry, pp. 1583-1586.

Malmsten, M. (2002). Surfactants and Polymers in Drug Delivery. pp. 1-17. Marcel Dekker Incorporated, ISBN: 0-8247-0804-0, New York, USA.

Michel, J.C.; Riviere, L.M.; Bellon-Fontaine, M.N. & Allaire, C. (1997). Effects of Wetting agents on the wettability of air-dried Sphagnum peats. Peat in Horticulture 2-7 Nov, 74-79.

Milks, R.; Fonteno, W. & Larson, R. (1989). Hydrology of horticultural substrates, III. Predicting air and water content of limited-volume plug cells. Journal of the American Society for Horticultural Science 114(1), 57-61.

Ortega, M.C.; Moreno, M.T.; Ordovás, J. & Aguado, M.T. (1996). Behaviour of different horticultural species in phytotoxicity bioassays of bark substrates.Scientia Horticulturae 66, 125-132.

Otten, W.; Raats, P.A.C.; Challa, H. & Kabat, P. (1999). Spacial and temporal dynamics of water in the root environment of potted plants on a flooded bench fertigation system. Netherlands Journal of Agricultural Science 47, 51-65

Plaut, Z.; Zieslin, N. & Arnon, Y. (1973). The influence of moisture regime on greenhouse rose production in various growth media. Scientia Horticulturae 1, 239-250.

Plaut, Z. & Zieslin, N. (1974). Productivity of greeenhouse roses following changes in soil moisture and soil air regimes. Scientia Horticulturae 2, 137-143.

Powell, D. (1986). Wetting agents Tools to control water movement. Ohio Florists Assosiation Bulletin 681, 6-8.

RAE (Real Academia Española), (2001). Diccionario de la Lengua Española, 22ª edición. La Real Academia Española, Madrid.

Raviv, M.;Wallach, R.; Silber, A. & Bar-Tal, A. (2002). Substrates and their analysis. in: Savvas, D., Passam, H. (Eds.), Hydroponic Production of Vegetables and Ornamentals, Embryo Publications, ISBN 960-8002-12-5, Athens, Greece.

Reinikainen, O. & Herranen, M. (1997). The influence of wetting agent on physical properties of peat. Acta Horticulturae 450, 375-379.

Riviere, L.M. & Nicolas, H. (1987). Conduite de Irrigation des cultures hors sol sur substrants, contraintes liees au choix des substrats. Milieux poreux et transferts hydriques. Bull GFHN 22, 47-70.

Riviere, L.M. & Caron, J. (2001). Research on substrates: state of the art and need for the coming 10 years. Acta Horticulturae 548, 29-14.

Rosen, M.J. (2004). Surfactants and interfacial phenomena, 3ª ed., John Wiley and Sons, Incorporated, pp, 1-33. ISBN 0471478180, Hoboken, NJ, USA.

Sahin, U.; Anapali, O. & Ercisli, S. (2002). Physico-chemical and physical properties of some substrates used in horticultura.Gartenbauwissenschaft 67(2), 55-60.

Salas, M.C. & Urrestarazu, M. (2004). Métodos de riego y fertirrigación en cultivo sin suelo. In: Urrestarazu, M., Tratado de cultivo sin suelo, 3ª edición, Ed. Mundi-Prensa, ISBN 84-8476-139-8, Madrid, España.

Shurtleff, M.C. & Averre, C.W. (1998). Glossary of plant-pathological terms. APS Press, pp. 361. ISBN 0-89054-176-0, Minesota, USA.

Tadros, T.F. (2005). Applied surfactants, principles and applications. Wiley- VCH Verlag GmbH & Co. KGaA, pp. 634. ISBN: 3-527-30629-3, Weinheimen, Germany.

Thacker, J.R. (2003). Pesticide adjuvants.Enciclopedia of Agrochemicals, pp. 1199-1205.

Urrestarazu, M. (2004), Tratado de cultivos sin suelo, 3ª ed. Mundi-Prensa, Servicio de Publicaciones de la Universidad de Almería, Madrid, España, pp. 914. ISBN 84-8476-139-8,, Madrid, Spain.

Urrestarazu, M. & Mazuela, P.C. (2005a). Effect of slow-release oxygen suplí by fertigation on horticultural crops under soilless culture. ScientiaHorticulturae 106, 484-490.

Urrestarazu, M. & Mazuela, P.C. (2005b). La pujanza de los cultivos sin suelo en la horticultura protegida. Vida Rural 205, 25-28.

Urrestarazu, M. & Guillén, C. (2006). Application of wetting agent by fertigation in horticultural crops on substrate. Tesis de Doctorado. Universidad de Almería, Spain.

Urrestarazu, M.; Guillén, C.; Mazuela, P.C. & Carrasco, G. (2008). Wetting agent effect on physical properties on new and reused rockwool and coconut coir waste. Scientia Horticulturae, 116 (1) 104-108.

Wilson, G. (1985), Effects of additives to peat on the air and water capacity, Acta Horticulturae 172, 207-209.

WSSA Herbicide Handbook, 7th ed. (1994). Champaign, IL, Weed Science Society of America.pp. 313.

Young, B.G. (2003). Herbicide adjuvants. Enciclopedia of Agrochemicals, pp. 707-718.

Zucconi, F.; Forte, M. & De Bertoldo, M. (1981). Biological evaluation of compost maturity.BioCycle (July/August), 27-29.

Permissions

The contributors of this book come from diverse backgrounds, making this book a truly international effort. This book will bring forth new frontiers with its revolutionizing research information and detailed analysis of the nascent developments around the world.

We would like to thank Dr. Alejandro Isabel Luna Maldonado, for lending his expertise to make the book truly unique. He has played a crucial role in the development of this book. Without his invaluable contribution this book wouldn't have been possible. He has made vital efforts to compile up to date information on the varied aspects of this subject to make this book a valuable addition to the collection of many professionals and students.

This book was conceptualized with the vision of imparting up-to-date information and advanced data in this field. To ensure the same, a matchless editorial board was set up. Every individual on the board went through rigorous rounds of assessment to prove their worth. After which they invested a large part of their time researching and compiling the most relevant data for our readers. Conferences and sessions were held from time to time between the editorial board and the contributing authors to present the data in the most comprehensible form. The editorial team has worked tirelessly to provide valuable and valid information to help people across the globe.

Every chapter published in this book has been scrutinized by our experts. Their significance has been extensively debated. The topics covered herein carry significant findings which will fuel the growth of the discipline. They may even be implemented as practical applications or may be referred to as a beginning point for another development. Chapters in this book were first published by InTech; hereby published with permission under the Creative Commons Attribution License or equivalent.

The editorial board has been involved in producing this book since its inception. They have spent rigorous hours researching and exploring the diverse topics which have resulted in the successful publishing of this book. They have passed on their knowledge of decades through this book. To expedite this challenging task, the publisher supported the team at every step. A small team of assistant editors was also appointed to further simplify the editing procedure and attain best results for the readers.

Our editorial team has been hand-picked from every corner of the world. Their multi-ethnicity adds dynamic inputs to the discussions which result in innovative outcomes. These outcomes are then further discussed with the researchers and contributors who give their valuable feedback and opinion regarding the same. The feedback is then collaborated with the researches and they are edited in a comprehensive manner to aid the understanding of the subject.

Apart from the editorial board, the designing team has also invested a significant amount of their time in understanding the subject and creating the most relevant covers. They scrutinized every image to scout for the most suitable representation of the subject and create an appropriate cover for the book.

The publishing team has been involved in this book since its early stages. They were actively engaged in every process, be it collecting the data, connecting with the contributors or procuring relevant information. The team has been an ardent support to the editorial, designing and production team. Their endless efforts to recruit the best for this project, has resulted in the accomplishment of this book. They are a veteran in the field of academics and their pool of knowledge is as vast as their experience in printing. Their expertise and guidance has proved useful at every step. Their uncompromising quality standards have made this book an exceptional effort. Their encouragement from time to time has been an inspiration for everyone.

The publisher and the editorial board hope that this book will prove to be a valuable piece of knowledge for researchers, students, practitioners and scholars across the globe.

List of Contributors

Ona Bundinienė, Danguolė Kavaliauskaitė, Roma Starkutė, Julė Jankauskienė, Vytautas Zalatorius and Česlovas Bobinas
Institute of Horticulture, Lithuanian Research Centre for Agriculture and Forestry, Lithuania

N. Bettahar
Laboratory Water & Environement, Department of Hydraulic, University Hassiba Ben Bouali, Chlef, Algeria

Ligia Gabriela Gheţea and Rozalia Magda Motoc
University of Bucharest, Faculty of Biology, Bucharest, Romania

Carmen Florentina Popescu
National Institute of Research & Development for Biotechnologies in Horticulture, Ştefăneşti-Argeş, Romania

Nicolae Barbacar, Tatiana Bătrînu, Ina Bivol and Ioan Baca
Genetics and Plant Physiology Institute of the Moldavian Academy of Sciences, Chişinău, Republic of Moldova

Ligia Elena Bărbării, Carmen Monica Constantinescu and Daniela Iancu
National Institute of Legal Medicine "Mina Minovici", Bucharest, Romania

Gheorghe Savin
National Institute for Viticulture and Oenology, Department of Grapevine Genetic Resources, Chişinău,
Republic of Moldova

Humberto Rodriguez-Fuentes, Juan Antonio Vidales-Contreras and Alejandro Isabel Luna-Maldonado
Department of Agricultural and Food Engineering, Faculty of Agriculture, Autonomous University of Nuevo Leon, Escobedo, Nuevo Leon, Mexico

Juan Carlos Rodriguez-Ortiz
Faculty of Agriculture, Autonomous University of San Luis Potosi, San Luis Potosi, Mexico

Pranas Viskelis, Ramune Bobinaite, Marina Rubinskiene, Audrius Sasnauskas and Juozas Lanauskas
Institute of Horticulture, Lithuanian Research Centre for Agriculture and Forestry, Lithuania

Albert Ayorinde Abegunde
Department of Urban and Regional Planning, Obafemi Awolwo University, Ile Ife, Nigeria

María Lelia Pochettino, Julio A. Hurrell and Verónica S. Lema
Laboratorio de Etnobotánica y Botánica Aplicada, Facultad de Ciencias Naturales y Museo, Universidad Nacional de La Plata, República Argentina
Consejo Nacional de Investigaciones Científicas y Técnicas (CONICET), República Argentina

Alejandro Isabel Luna-Maldonado
Department of Agricultural and Food Engineering, Faculty of Agriculture, Autonomous University of Nuevo Leon, Escobedo, N.L., Mexico

Clement Vigneault
Department of Bioresource Engineering, Macdonald Campus, McGill University, Ste-Anne-de-Bellevue, Québec, Canada

Kei Nakaji
Kyushu University, Laboratory of Agricultural Ecology, Division of Agricultural Ecology, Department of Plant Resources, Fukuoka, Japan

Carlos Guillén and Miguel Urrestarazu
Universidad of Almería, Spain

Printed in the USA
CPSIA information can be obtained
at www.ICGtesting.com
JSHW011350221024
72173JS00003B/250